Lab Manual to Accompany Modern Livestock and Poultry Production

Cengage Learning® is proud to support FFA activities

Lab Manual to Accompany Modern Livestock and Poultry Production

9th Edition

Frank B. Flanders
James R. Gillespie
Levi Cahan

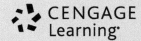

Australia • Brazil • Japan • Korea • Mexico • Singapore • Spain • United Kingdom • United States

Lab Manual to Accompany Modern Livestock and Poultry Production, 9th Edition
Frank B. Flanders, James R. Gillespie & Levi Cahan

SVP, GM Skills & Global Product Management: Dawn Gerrain

Product Director: Matthew Seeley

Product Team Manager: Erin Brennan

Product Manager: Nicole Sgueglia

Senior Director, Development: Marah Bellegarde

Senior Product Development Manager: Larry Main

Senior Content Developer: Jennifer Starr

Content Developer: Richard Hall

Product Assistant: Jason D. Koumourdas

Vice President, Marketing Services: Jennifer Ann Baker

Marketing Manager: Scott Chrysler

Senior Production Director: Wendy Troeger

Production Director: Andrew Crouth

Senior Content Project Manager: Elizabeth C. Hough

Production Service: MPS Limited

Senior Art Director: Benjamin Gleeksman

Digital Project Manager: Christina Brown

Cove/Design images: Hay field background image: ©Mikael Goransson/Shutterstock.com; hereford cow: ©iStockphoto/John Nielsen; two piglets:©iStockphoto/Jeff Fullerton; feeding goats:©iStockphoto/BORSEV; Horse: ©iStockphoto/Cathleen Abers-Kimball; hens: ©iStockphoto/johnnyscriv; dairy cow: ©iStockphoto/Jason Lugo

© 2016 Cengage Learning

WCN: 01-100-101

ALL RIGHTS RESERVED. No part of this work covered by the copyright herein may be reproduced, transmitted, stored, or used in any form or by any means graphic, electronic, or mechanical, including but not limited to photocopying, recording, scanning, digitizing, taping, Web distribution, information networks, or information storage and retrieval systems, except as permitted under Section 107 or 108 of the 1976 United States Copyright Act, without the prior written permission of the publisher.

> For product information and technology assistance, contact us at
> **Cengage Learning Customer & Sales Support, 1-800-354-9706**
> For permission to use material from this text or product,
> submit all requests online at **www.cengage.com/permissions**.
> Further permissions questions can be e-mailed to
> **permissionrequest@cengage.com**

Library of Congress Control Number: 2014939094

ISBN: 978-1-133-28354-6

Cengage Learning
20 Channel Center Street
Boston, MA 02210
USA

Cengage Learning is a leading provider of customized learning solutions with office locations around the globe, including Singapore, the United Kingdom, Australia, Mexico, Brazil, and Japan. Locate your local office at **www.cengage.com/global**

Cengage Learning products are represented in Canada by Nelson Education, Ltd.

To learn more about Cengage Learning, visit **www.cengage.com**

Purchase any of our products at your local college store or at our preferred online store **www.cengagebrain.com**

Notice to the Reader

Publisher does not warrant or guarantee any of the products described herein or perform any independent analysis in connection with any of the product information contained herein. Publisher does not assume, and expressly disclaims, any obligation to obtain and include information other than that provided to it by the manufacturer. The reader is expressly warned to consider and adopt all safety precautions that might be indicated by the activities described herein and to avoid all potential hazards. By following the instructions contained herein, the reader willingly assumes all risks in connection with such instructions. The publisher makes no representations or warranties of any kind, including but not limited to, the warranties of fitness for particular purpose or merchantability, nor are any such representations implied with respect to the material set forth herein, and the publisher takes no responsibility with respect to such material. The publisher shall not be liable for any special, consequential, or exemplary damages resulting, in whole or part, from the readers' use of, or reliance upon, this material.

Printed at CLDPC, USA, 02-20

Contents

PREFACE   xi

ABOUT THE AUTHOR   xi

SECTION 1   **THE LIVESTOCK INDUSTRY**

- **Chapter 1**   Domestication and Importance of Livestock   2
  - Introduction   2
  - 1-1 Domestication Timeline Activity   3
  - 1-2 Class Activity: Leading States   4
  - 1-3 Trend Analysis Activity   5
  - 1-4 Food Safety   7
  - Matching Activity   9
  - Lab Questions   10
- **Chapter 2**   Career Opportunities in Animal Science   11
  - Introduction   11
  - 2-1 Employment Opportunities Assessment   12
  - 2-2 Career Exploration Activity   15
  - 2-3 Job Interview Preparation   17
  - 2-4 Fill-in-the-Blank Activity   19
  - Lab Questions   20
- **Chapter 3**   Safety in Livestock Production   21
  - Introduction   21
  - Hazards in Livestock Handling   22
  - 3-1 Facilities Design with Safety in Mind   23
  - 3-2 Livestock Handling Injury Report   25
  - Lab Questions   30
- **Chapter 4**   Livestock and the Environment   31
  - Introduction   31
  - 4-1 Environmental Impact of Wastes (Federal and State Laws)   33
  - 4-2 Manure Application Calculations   34
  - Matching   36
  - Lab Questions   37

SECTION 2   **ANATOMY, PHYSIOLOGY, FEEDING, AND NUTRITION**

- **Chapter 5**   Anatomy, Physiology, and Absorption of Nutrients   40
  - Introduction   40
  - 5-1 Anatomy Model Lab   42
  - 5-2 Functions of Hormones   46
  - Matching Activity   49
  - Lab Questions   50
- **Chapter 6**   Feed Nutrients   51
  - Introduction   51
  - 6-1 Feed Sampling Activity   52
  - 6-2 Grain Elevator Map   57
  - Matching   59
  - Lab Questions   60
- **Chapter 7**   Feed Additive and Growth Promotants   61
  - Introduction   61
  - 7-1 Feed Additive Analysis   62
  - 7-2 Antibiotic Residues   64
  - Matching Activity   68
  - Lab Questions   69

Chapter 8	Balancing Rations	70
	• Introduction	70
	• 8-1 Dry-Matter Calculation	72
	• 8-2 Balancing a Ration Using Equations	75
	• Matching Activity	78
	• Lab Questions	79

SECTION 3 — ANIMAL BREEDING

Chapter 9	Genetics of Animal Breeding	82
	• Introduction	82
	• 9-1 Punnett Square	83
	• 9-2 Dominant vs. Recessive Allele Activity	87
	• Matching Activity	88
	• Lab Questions	89
Chapter 10	Animal Reproduction	90
	• Introduction	90
	• 10-1 Reproductive Planning for Livestock	91
	• 10-2 Reproductive Anatomy Activity	93
	• Matching Activity	96
	• Lab Questions	97
Chapter 11	Biotechnology in Livestock Production	98
	• Introduction	98
	• 11-1 Impacts of Using rbST	100
	• 11-2 Genetically Modified Organisms Debate Activity	101
	• Matching Activity	103
	• Lab Questions	104
Chapter 12	Animal Breeding Systems	105
	• Introduction	105
	• 12-1 Improving Herd Quality through Proper Breeding Strategy	107
	• Matching	109
	• Lab Questions	110

SECTION 4 — BEEF CATTLE

Chapter 13	Breeds of Beef Cattle	112
	• Introduction	112
	• 13-1 Breed Selection Activity	113
	• Matching Activity	115
	• Lab Questions	116
Chapter 14	Selection and Judging of Beef Cattle	117
	• Introduction	117
	• 14-1 Body Condition Scoring Activity	118
	• 14-2 Giving Oral Reasons	121
	• Matching Activity	122
	• Lab Questions	123
Chapter 15	Feeding and Management of the Cow-Calf Herd	124
	• Introduction	124
	• 15-1 Feeding Cows and Calves at Different Stages of Growth	126
	• 15-2 Calf Care Activity	128
	• Matching Activity	130
	• Lab Questions	131
Chapter 16	Feeding and Management of Feeder Cattle	132
	• Introduction	132
	• 16-1 Feeder Cattle Calculation Lab	134
	• Matching Activity	136
	• Lab Questions	137

Chapter 17	Diseases and Parasites of Beef Cattle	138
	• Introduction	138
	• 17-1 Disease Prevention and Treatment Activity	139
	• Lab Questions	140
Chapter 18	Beef Cattle Housing and Equipment	141
	• Introduction	141
	• 18-1 Design a Modern Cattle-Handling Facility Lab	143
	• 18-2 Equipment Cost Lab	146
	• Matching Activity	148
	• Lab Questions	149
Chapter 19	Marketing Beef Cattle	150
	• Introduction	150
	• 19-1 Marketing Methods Lab	152
	• 19-2 Alternative Beef Markets Activity	153
	• Matching Activity	154
	• Lab Questions	155

SECTION 5   SWINE

Chapter 20	Breeds of Swine	158
	• Introduction	158
	• 20-1 Breed History and Function Activity	160
	• Matching Activity	163
	• Lab Questions	164
Chapter 21	Selection and Judging of Swine	165
	• Introduction	165
	• 21-1 Using Performance Data in Swine Selection	167
	• Matching Activity	171
	• Lab Questions	172
Chapter 22	Feeding and Management of Swine	173
	• Introduction	173
	• 22-1 Feed Impacts on Gestation Lab	175
	• 22-2 Farrowing Practices Activity	176
	• Matching Activity	177
	• Lab Questions	178
Chapter 23	Diseases and Parasites of Swine	179
	• Introduction	179
	• 23-1 Diseases That Affect Swine	181
	• Matching Activity	184
	• Lab Questions	185
Chapter 24	Swine Housing and Equipment	186
	• Introduction	186
	• 24-1 Equipment Inventory Worksheet	188
	• Matching Activity	190
	• Lab Questions	191
Chapter 25	Marketing Swine	192
	• Introduction	192
	• 25-1 Marketing Methods Lab	193
	• Matching Activity	194
	• Lab Questions	195

SECTION 6 — SHEEP AND GOATS

Chapter 26	Breeds and Selection of Sheep	198
	• Introduction	198
	• 26-1 Sheep Breed Identification	199
	• 26-2 Market Lamb Lab	201
	• Matching Activity	202
	• Lab Questions	203
Chapter 27	Feeding, Management, and Housing of Sheep	204
	• Introduction	204
	• 27-1 Feeding Considerations	206
	• 27-2 Crossbreeding Activity	208
	• Matching Activity	209
	• Lab Questions	210
Chapter 28	Breeds, Selection, Feeding, and Management of Goats	211
	• Introduction	211
	• 28-1 Goat Breed Identification	213
	• 28-2 Design a Goat Facilities Lab	214
	• Matching Activity	215
	• Lab Questions	216
Chapter 29	Diseases and Parasites of Sheep and Goats	217
	• Introduction	217
	• 29-1 Tapeworm lab	218
	• 29-2 Parasite Identification Lab	219
	• Matching Activity	220
	• Lab Questions	221
Chapter 30	Marketing Sheep, Goats, Wool, and Mohair	222
	• Introduction	222
	• 30-1 Classing and Grading of Sheep	224
	• 30-2 Marketing Wool	225
	• Matching Activity	226
	• Lab Questions	227

SECTION 7 — HORSES

Chapter 31	Selection of Horses	230
	• Introduction	230
	• 31-1 Breed Selection	231
	• 31-2 Anatomy Identification Exercise	234
	• 31-3 Unsoundness Activity	235
	• Matching Activity	237
	• Lab Questions	238
Chapter 32	Feeding, Management, Housing, and Tack	239
	• Introduction	239
	• 32-1 Feeding for Appropriate Size and Age	240
	• 32-2 Horse Tack Identification	242
	• Matching Activity	245
	• Lab Questions	246
Chapter 33	Diseases and Parasites of Horses	247
	• Introduction	247
	• 33-1 Equine Disorders	248
	• 33-2 Parasite Life Cycle Lab	249
	• Matching Activity	253
	• Lab Questions	254

Chapter 34	Training and Horsemanship	**255**
	• Introduction	**255**
	• 34-1 Training a Young Horse Activity	**257**
	• 34-2 Equine Performance Events	**259**
	• Matching Activity	**260**
	• Lab Questions	**261**

SECTION 8   POULTRY

Chapter 35	Selection of Poultry	**264**
	• Introduction	**264**
	• 35-1 Growth of the Poultry Industry	**265**
	• 35-2 Chicken Anatomy Activity	**266**
	• Matching Activity	**268**
	• Lab Questions	**269**
Chapter 36	Feeding, Management, Housing, and Equipment	**270**
	• Introduction	**270**
	• 36-1 Water Consumption Lab	**271**
	• 36-2 Management Lab: Broilers vs. Layers	**274**
	• Matching Activity	**275**
	• Lab Questions	**276**
Chapter 37	Diseases and Parasites of Poultry	**277**
	• Introduction	**277**
	• 37-1 Disease and Parasite Control Program Activity	**278**
	• Matching Activity	**281**
	• Lab Questions	**282**
Chapter 38	Marketing Poultry and Eggs	**283**
	• Introduction	**283**
	• 38-1 Egg Grading Lab	**285**
	• Matching Activity	**288**
	• Lab Questions	**289**

SECTION 9   DAIRY CATTLE

Chapter 39	Breeds of Dairy Cattle	**292**
	• Introduction	**292**
	• 39-1 Trends in Dairy Production Activity	**293**
	• 39-2 Breed Comparison Lab	**295**
	• Matching Activity	**297**
	• Lab Questions	**298**
Chapter 40	Selecting and Judging Dairy Cattle	**299**
	• Introduction	**299**
	• 40-1 Dairy Cow Scorecard	**300**
	• Matching Activity	**302**
	• Lab Questions	**303**
Chapter 41	Feeding Dairy Cattle	**304**
	• Introduction	**304**
	• 41-1 Pearson's Square	**305**
	• 41-2 Total Mixed Ration Lab	**306**
	• Matching Activity	**309**
	• Lab Questions	**310**

Chapter 42	Management of the Dairy Herd	311
	• Introduction	311
	• 42-1 Dairy herd record keeping activity	313
	• 42-2 Cull Cow Selection Lab	314
	• Matching Activity	315
	• Lab Questions	316
Chapter 43	Milking Management	317
	• Introduction	317
	• 43-1 Milk Quality Investigation	318
	• Matching Activity	320
	• Lab Questions	321
Chapter 44	Dairy Herd Health	322
	• Introduction	322
	• 44-1 Develop a Herd Health Plan Lab	323
	• Matching Activity	326
	• Lab Questions	327
Chapter 45	Dairy Housing and Equipment	328
	• Introduction	328
	• 45-1 Design a Modern Cattle Handling Facility Lab	330
	• Lab Questions	332
Chapter 46	Marketing Milk	333
	• Introduction	333
	• 46-1 Dairy Advertising Activity	334
	• 46-2 Milk Grading Lab	336
	• Lab Questions	338

SECTION 10 ALTERNATIVE ANIMALS

Chapter 47	Rabbits	340
	• Introduction	340
	• 47-1 Breed Selection Activity	342
	• 47-2 Rabbit Health Problem Lab	344
	• Matching Activity	346
	• Lab Questions	347
Chapter 48	Bison, Ratites, Llamas, Alpacas, and Elk	348
	• Introduction	348
	• 48-1 Bison Management Lab	350
	• 48-2 History of Llamas and Alpacas Activity	352
	• Matching Activity	353
	• Lab Questions	354

REFERENCES 355

Preface

The *Lab Manual* to accompany *Modern Livestock and Poultry Production*, Ninth Edition, is designed for agricultural education students who require competency in all phases and types of livestock production. The content contained in the *Lab Manual* is based on the most up-to-date information available and is applicable to all areas of the United States. This *Lab Manual*, complete with exercises and activities, accompanies the critical content areas covered throughout the book. The *Lab Manual* reinforces text content through the practical application of real-world examples. It is recommended that students complete corresponding lab components to further develop understanding of essential skills and concepts discussed in the text.

About the Author

Levi Cahan has a distinguished background in livestock production from years of training and educating others as well as being self-employed as a farmer. He is the lead Agriculture Educator at Schuylerville High School in upstate New York, where he instructs and manages student learning in several agricultural subjects with a focus on animal science. He received his BS in Animal Science and his MS in Agricultural Education from Cornell University. He also studied abroad in New Zealand at Lincoln University, specializing in animal science and rotational grazing practices. While continuing to teach agriculture to high school students, Mr. Cahan has also built an agricultural business from the ground up. His farm currently raises livestock using all natural and alternative practices. After raising grass-fed beef, pork, and poultry animals, he retails the specialty meat products directly to farmers' markets in New York City. This provides another avenue for him to educate and share with the public. Mr. Cahan stays active in agriculture and education as an FFA advisor; he has served on the New York State FFA Governing Board as a trustee, as a trustee for the NYAAE, as a trustee on the NYS FFA Foundation board, and as a member of the NYS beef council.

SECTION 1

The Livestock Industry

Chapter 1	Domestication and Importance of Livestock	2
Chapter 2	Career Opportunities in Animal Science	11
Chapter 3	Safety in Livestock Production	21
Chapter 4	Livestock and the Environment	31

Chapter 1

Domestication and Importance of Livestock

INTRODUCTION

The domestication of animals played a vital role in the development of civilization. Domesticated animals provided a dependable source of food and clothing. Most of the ancestors of present-day farm animals were first tamed in the Neolithic (New Stone) Age. Early explorers brought the various species of farm animals to the United States. These animals spread across the United States as early explorers and colonists moved westward. Certain states became leaders in livestock production due to factors such as climate, transportation routes, and access to natural resources.

Animals have many useful functions. They convert grain feeds into other food products, are a source of materials for clothing, and provide power and recreation. Animals concentrate bulky feeds, making them easier to market. Many by-products of animals are also important to society. Consumer concerns affect the trends in the livestock industries. Understanding the market inventory trends as well as consumer preferences is important for livestock business owners.

The most important consumer concern is the safety of our food products. The United States has some of the highest food safety regulations in the world. The U.S. Department of Agriculture (USDA) and the Food and Drug Administration (FDA) are responsible for creating and enforcing food safety laws. Several environmental factors can cause conditions that allow bacteria to grow on the food. According to the Centers for Disease Control and Prevention (CDC), the four major bacterial pathogens that contaminate meat and poultry products are *Salmonella*, *Campylobacter*, *E. coli* O157:H7, and *Listeria monocytogenes*.

The following lab exercises will focus on developing an understanding of the history of animal domestication, analyzing market trends, and maintaining a safe supply of livestock products.

Name_____ Date_____

EXERCISE 1-1 DOMESTICATION TIMELINE ACTIVITY

For each of the animals commonly domesticated in the United States, fill in the blank squares on the timeline chart with proper information about the history of domestication for each species. You will need to include information on (1) classification, (2) the place and period of time each animal was first domesticated by humans, (3) major dates in the history of domestication in the United States, and (4) the key functions and uses of the animal by humans.

Domestication Timeline

Common Name	Family Genus Species	Time and Place of Origin	Arrival in the United States/Other Historical Dates	Function of Animal in Agriculture
Cattle				
Swine				
Sheep				
Goats				
Horses				
Poultry				

© 2016 Cengage Learning®. May not be scanned, copied or duplicated, or posted to a publicly accessible website, in whole or in part.

Name_____ Date_____

EXERCISE 1-2 CLASS ACTIVITY: LEADING STATES

Use Figure 1-4 from the textbook to compare livestock production among states within the United States. The class will need to be divided into groups and each group assigned a category of livestock production. Each group should research the top three producing states from their category. They will need to give a presentation that explains why these particular states are suited to produce this livestock category. Using Internet resources, research the agricultural production statistics for the state, as well as other characteristics that may affect the ability to produce livestock products.

The following explains the criteria for the presentation:

Format: Each group's presentation will be given to the class using oral and visual content. The visual component may be developed using computer presentation programs, or using physically prepared posters.

Content: Each state should be described, highlighting characteristics that make it a good location to produce the livestock category. Examples of characteristics may be:

- geological attributes
- natural resources
- historical factors
- consumer demands
- transportation routes
- climatic factors
- population size and characteristics

Name_____ Date_____

EXERCISE 1-3 TREND ANALYSIS ACTIVITY

The growth of livestock markets, and the decrease in the number of farms has encouraged nonagricultural investors to seek out agricultural economic opportunities. Investors without agricultural experience rely on consultants for direction and instruction on how to invest their money.

For this exercise, you will take on the role of an agricultural business consultant. For Part 1, your task will be to analyze the data given in Figures 1-5 and 1-6 of the textbook. Using the analyzed data, in Part 2, you will prepare a proposal for your investor that highlights the trends in each livestock market and suggests which livestock category to invest in.

Both figures show the long-term projection numbers from the USDA. Using these data, create a line graph that shows the projected growth and decline of livestock populations from 1995 through 2020 (a graph template has been provided for you). Label the lines so it is easy to determine which animal each line represents. Graph 1 should be used to record the data from Figure 1-5, and graph 2 should contain the data from Figure 1-6.

Part 1: Complete the graphs using the appropriate data.

Graph # 1:

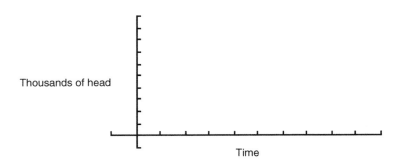

Graph # 2:

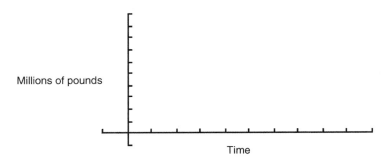

Part 2: For each of the livestock categories, describe the current market trends and projections for the future.

Beef Cattle Trends:

Current inventory numbers _____

Projected numbers for 2020 _____

Market factors _____

Name _____ Date _____

Dairy Trends:
Current inventory numbers _____
Projected numbers for 2020 _____
Market factors _____

Hog Trends:
Current inventory numbers _____
Projected numbers for 2020 _____
Market factors _____

Chicken Trends:
Current inventory numbers _____
Projected numbers for 2020 _____
Market factors _____

Turkey Trends:
Current inventory numbers _____
Projected numbers for 2020 _____
Market factors _____

What is your final suggested livestock market?
_____
_____
_____
_____

Give your reasons for the choice.
_____
_____
_____
_____
_____

Name_____ Date_____

EXERCISE 1-4　FOOD SAFETY

Foodborne illness is a serious concern facing all animal product markets. In recent years, the United States has improved its standards for sanitation and prevention of foodborne bacteria. All USDA-certified meat-processing plants in the United States must adhere to federal sanitation regulations and must operate with the presence of a USDA inspector. The system known as FAT-TOM (Figure 1-16) is used as a guide to understand environmental conditions that allow optimum growth of bacteria so that these conditions can be avoided.

Since 2000, all large meat-processing facilities have been required to comply with Hazard Analysis and Critical Control Point (HACCP) standards. There are seven principles that slaughterhouse in the United States must adhere to. The seven HACCP principles can be found in the textbook.

Food Safety Scenario

In this scenario, you, the student, will take the role of a USDA slaughterhouse inspector. Your job is to use the data that have been collected during a recent hazard analysis of a local meat-processing facility to apply the HACCP principles. Use the FAT-TOM guidelines for reference as you conduct a review of the slaughter plant. If any violations are found, you must give recommendations for how the plant can become compliant as well as a timeline for making the changes.

The plant that is being reviewed has had some compliance issues in the past, so frequent inspections and strict enforcement of HACCP standards is necessary. During the most recent hazard analysis, the following environmental conditions were recorded:

> The hazard analysis was conducted on June 27; the temperature was 86 degrees. In the processing facilities, the temperature inside of the plant's walk-in cooler is normally 36 degrees. This temperature is maintained to preserve the carcasses as they hang prior to processing and packaging. On that morning, the compressor of the cooler malfunctioned, and it took 10 hours for it to be repaired. The cooler maintained a temperature below 41 degrees for 5 hours, after which the temperature entered the "danger zone." While this repair was being conducted a puddle of stagnant water with a "wa" of 0.56 was found in the ceiling of the cooler; it had been leaking onto the carcasses as they hung. It was also found that a normal practice around the plant was to leave the cooler door open for hours at a time, letting outside air into the cooler. Another finding of concern was the time it would take the employees from the moment a beef carcass came out of the cooler until it was completely packaged in airtight packaging. That time was at least 4 hours. Another concern was the presence of stagnant water in several places throughout the facility. After samples were tested, the presence of salmonella and E. coli were found to be above the critical limits.

Food Safety Analysis

As the USDA inspector, create a report that identifies the violations of the HACCP guidelines as well as the steps and controls to ensure that the slaughter house can become compliant with the rules and avoid being shut down. Complete a

Name_____ Date_____

HACCP analysis for the local slaughterhouse, using the seven principles discussed in the textbook (for example, schedule periodic testing, establish quarterly reviews).

1. Findings of hazard analysis.
2. Identify critical control points.
3. Establish critical limits for CCPs.
4. Establish requirements for monitoring the critical controls.
5. Identify the corrective actions to be taken.
6. Establish record-keeping procedures for CCPs.
7. Establish procedures to verify the system is working.

Extension Activity

Share your finding with the class.

Name _____ Date _____

CHAPTER 1 MATCHING ACTIVITY

Directions: Match the term to the corresponding definition or description.

Term

____ 1. domesticate
____ 2. bos taurus
____ 3. bos indicus
____ 4. selection
____ 5. crossbreeding
____ 6. sus scrofa
____ 7. sus vittatus
____ 8. moufflons
____ 9. eohippus
____ 10. draft animal
____ 11. gallus gallus
____ 12. anas boschas
____ 13. non-ruminant
____ 14. ruminant
____ 15. insulin
____ 16. cortisone
____ 17. thrombin
____ 18. heparin
____ 19. epinephrine
____ 20. rennet
____ 21. corticotropin
____ 22. soring
____ 23. Hazard Analysis and Critical Control Points (HACCPs)
____ 24. critical control points (CCPs)
____ 25. food irradiation
____ 26. radura

Definition

a. extracted from the adrenal glands of animals and used for the treatment of hay fever and asthma

b. East Indian pig

c. animals, such as swine and poultry, that are fed large amounts of grain

d. to identify and breed animals with traits that are desirable by the breeder

e. animal used for pulling loads

f. extracted from the pituitary glands in the brain and used for the treatment of breathing problems, allergies, and leukemia

g. to adapt the behavior of an animal to fit the needs of people

h. extracted from the adrenal glands of animals and used for the treatment of rheumatoid arthritis

i. animals, such as cattle, sheep, and goats, that have a stomachs that are divided into many compartments

j. a tiny four-toed ancestor of the horse

k. wild big-horned sheep of Asia

l. wild mallard duck; ancestor of all domestic breeds of ducks

m. extracted from the pancreas of animals and used in the treatment of diabetes

n. the practice of using chemical or mechanical irritants on the four legs of a horse

o. modern cattle descendant; humped cattle found in tropical countries

p. wild jungle fowl of India; ancestor of most tame chickens

q. domestic cattle descending from the Aurochs or the Celtic Shorthorn

r. the point at which any step or procedure in the process can be applied to prevent, eliminate, or reduce a food safety hazard

s. developed for NASA to monitor the production of food for space programs

t. the treatment of food with radioactive isotopes to kill bacteria, insects, and colds that are present in the food

u. European wild boar

v. the international symbol all irradiated food must carry

w. extracted from the blood of animals and used as a coagulant during surgery to help blood clot

x. extracted from the stomachs of cattle and used to make cheese

y. extracted from the lungs of animals and used during surgery to prevent blood clots and heart attacks

z. mating of animals of different breeds

Name_____ Date_____

CHAPTER 1 LAB QUESTIONS

1. When evaluating hogs, beef, dairy, chickens, and turkeys, which market will have the most growth potential? Which market is second? Third?

_____
_____
_____
_____
_____
_____

2. From the list of livestock species discussed in this chapter, which domesticated animal arrived in the United States first? How did this species arrive?

_____
_____
_____
_____
_____
_____

3. What are some common market factors that affect each livestock industry?

_____
_____
_____
_____
_____
_____

4. What are the six factors that determine if food is safe to eat and free of harmful bacteria?

_____
_____
_____
_____
_____

Chapter 2
Career Opportunities in Animal Science

INTRODUCTION

Careers relating to agriculture are more relevant than ever—more than 200 different careers are available to persons with an interest in agriculture. Many of the career paths in animal science are not directly related to raising animals; for example, there are a variety of careers available from genetic engineering to selling grain to farmers. In the past 25 years, the number of jobs in production agriculture has decreased along with the number of full-time farms in the United States. The trend shows that since 1936, the number of farms has decreased and the size of the farm has increased; this trend is expected to continue. The exception is a growing "locavore" movement, where small, niche farms produce products for local populations; consumers seek out those products to decrease the use of long-distance food transportation. Many independent restaurants are beginning to serve food prepared with all locally produced ingredients.

As the world's population grows rapidly, the demand for food is increasing as well. This demand is expanding the job market in careers that involve the processing, packaging, and marketing of food products. Many job opportunities in the animal industry are available in the fields of agribusiness, communications, science, government, education, sales, technology, engineering, and others. Previous experience working with animals is beneficial in any career path in the animal industry.

An increasing number of jobs that are available in animal science require a minimum education level of a bachelor's degree. In specialized areas of animal science, further education may be required such as a master's degree or a PhD. The number of job openings for college graduates with experience in agriculture over the next several years is expected to increase, while the number of qualified job applicants is not expected to keep up with the demand.

The objectives of the following exercises are to develop an understanding of the trends in the job market in the field of agriculture and to plan for the expected changes in employment opportunities in the future. You will conduct an employment opportunities assessment, research a career in animal agriculture, and prepare for an interview and job.

Name_____ Date_____

EXERCISE 2-1 EMPLOYMENT OPPORTUNITIES ASSESSMENT

Directions: For each career area given, list three potential job opportunities for graduates, and describe one of the occupations. Use Table 2-3 in the textbook for reference.

1. Career Area: **Scientists, Engineers, and Related Specialists**
   - Three examples of job opportunities:
   _____
   _____

   - Job description for one career:
   _____
   _____

2. Career Area: **Managers and Financial Specialists**
   - Three examples of job opportunities:
   _____
   _____

   - Job description for one career:
   _____
   _____

3. Career Area: **Marketing, Merchandising, and Sales Representatives**
   - Three examples of job opportunities:
   _____
   _____

   - Job description for one career:
   _____
   _____

4. Career Area: **Education, Communication, and Information Specialists**
   - Three examples of job opportunities:
   _____
   _____

   - Job description for one career:
   _____
   _____

Name_____ Date_____

5. Career Area: **Beef Production**
   - Three examples of job opportunities:
   _____
   _____

   - Job description for each career:
   _____
   _____

6. Career Area: **Dairy Occupations**
   - Three examples of job opportunities:
   _____
   _____

   - Job description for each career:
   _____
   _____

7. Career Area: **Swine Occupations**
   - Three examples of job opportunities:
   _____
   _____

   - Job description for each career:
   _____
   _____

8. Career Area: **Sheep and Goats Occupations**
   - Three examples of job opportunities:
   _____
   _____

   - Job description for each career:
   _____
   _____

9. Career Area: **Equine Occupations**
   - Three examples of job opportunities:
   _____
   _____

Name_____ Date_____

- Job description for each career:

  _____
  _____

10. Career Area: **Poultry Occupations**
    - Three examples of job opportunities:

      _____
      _____

    - Job description for each career:

      _____
      _____

Name_____ Date_____

EXERCISE 2-2 CAREER EXPLORATION ACTIVITY

After researching the careers available in the field of animal agriculture, choose one career in particular to research in more detail. Use the textbook as a resource, as well as the Internet, to find information that describes the career you have chosen.

Task: Using your research, answer the following questions.

1. Name of the career/job you researched
   _____
   _____

2. Area of animal science that the career fits into (livestock production, financial institutions, etc.)
   _____
   _____

3. Why did you select this career?
   _____
   _____

4. How/why is this career important in the production, marketing, or sales of animal products?
   _____
   _____

5. Skills and abilities needed to be successful in this career (paperwork, phone skills, people skills, lab skills, etc.)
   _____
   _____

6. Specialized skills required
   _____
   _____

7. Education needed in high school to pursue this career
   _____
   _____

8. What education is needed beyond high school?
   _____
   _____

Name_____ Date_____

9. What are the working conditions and situations that are present in this career?
_____
_____

10. Salary starting vs. experienced? Is it based on education level?
_____
_____

11. What are the advantages of this career?
_____
_____

12. What are the disadvantages of this career?
_____
_____

13. How can one find a job in this career?
_____
_____

14. Give three examples of current career opportunities in this field. Where are they located? What are the requirements of applicants?
_____
_____

15. Are you interested in pursuing this career? Why or why not?
_____
_____

Optional Exercise Extension Activity

Using the answers from questions 1–15, create a computer-based visual presentation on your chosen career. The presentation should be a minimum of eight slides, with a minimum of five visual aids. Use the answers to the questions for the content of the slides. Deliver a presentation to the class or a group of individuals who could be interested in this career field. The instructor will evaluate your presentation for accuracy and detail.

Name_____ Date_____

EXERCISE 2-3   JOB INTERVIEW PREPARATION

This exercise is modeled after the FFA Job Interviews Career Development Event. Your textbook, the Internet, and the FFA CDE handbook are all appropriate resources for your research.

Tasks

As an individual, you need to prepare materials prior to applying for the career you have chosen. For this assignment, you will prepare a resumé, a cover letter, a list of references, and a list of interview questions. You may also be responsible for a face-to-face mock interview with your teacher.

Note: For Parts 1, 2 and 3, use the same career that you described in Exercise 2-2.

Part 1: Resume

Your resume should consist of the following:

- Name and address
- Objectives
- Education
- Skills
- Experience
- Awards
- Extracurricular activities
- Hobbies
- Any other important personal information

(An example resumé is included on the next page for reference.)

Part 2: Cover Letter

Your cover letter should be addressed to a mock employer in your chosen career field. It should be typed, three paragraphs long, with your name and address in the heading. The letter should reference your skills and resumé information, as well as your desire to work in the specific field. You can mention key personal attributes and accomplishments.

Part 3: List of References

Prepare a list of names and contact information for at least three persons who would recommend you for this job. (They will not actually be contacted for this assignment.) Describe your relationship with each person. Make sure the relationship is appropriate for the job you are applying for.

Part 4: Interview

- Prepare a list of 10 interview questions that could be asked by an employer during an interview for this career.
- Prepare a list of 5 questions that would be appropriate to ask an employer about this career.

Name_____ Date_____

Jane Smith

123 Hilldale Ave

Anywhere Town, USA 12345

Phone: 555-242-5752; Cell: 555-545-4813

jsmith@cmail.com

Career Objective: To obtain a position as a veterinary assistant.

Education:

- Springfield High School, Anywhere Town, USA – Diploma, mixed curriculum – June 2014. Courses include introduction to agriculture, animal science, accounting, advanced biology, and chemistry.
- Accepted to Springfield Community College beginning in the fall.

Skills: cared for animals and children, frequently working with the public, responsible for placing supply orders and managing receipt of material, and inventory update.

Experience:

Kennel Attendant – 2010 – Present

- Walked dogs, cleaned kennels, and provided customer service.

Child Care – 2009 – Present

- Provide childcare to neighbor children in the summer.

Achievements:

State FFA Degree Recipient

State Finalist in the Small Animal Care Career Development Event

Honor Roll – 2010, 2012, 2013

Volunteer Experience:

Vineyard Food Pantry – 2012 – present

SPCA – 2012 – present

Extra Curricular Activities:

FFA Leadership Organization – 2008 – present

High School Band – 2011–2014

Youth Group Member – 2012 – present

Varsity Track and Field – 2010–2013

References:

Jim Smith	Agriculture Teacher/FFA Advisor	555-454-4455
Haley Sprout	Employer, Sprout Kennels	555-454-3465
Tyler Kent	Youth Group Leader	555-454-6367

Name_____ Date_____

EXERCISE 2-4 FILL-IN-THE-BLANK ACTIVITY

Directions: Use the word bank below to fill in the blanks in the following statements.

Word Bank:

associate's degree   bachelor's degree   interest

doctoral degree   talent   citizenship

master's degree

1. In order to become certified to teach animal science courses at any high school in the country, he went to graduate school and received his _____.

2. A local beef farmer showed _____ by volunteering a day this summer at an annual local farm tour, giving hay rides to visitors.

3. Before Josh got hired by a dairy equipment company to install equipment into milking parlors, he attended a local college for two years and received his _____.

4. The veterinarian has developed a _____ for successfully administering injections to animals quickly and completely.

5. She earned her _____ in farm business management after attending a state university for four years.

6. The teenager developed an _____ in horses after he saw the stable manager help a mare give birth to a healthy foal.

7. A veterinarian has to be successful through the challenges of vet school and residency in order to get his or her _____.

Name_____ Date_____

CHAPTER 2 LAB QUESTIONS

1. Describe the occupation trends in agriculture and animal science over the past 25 years.

   _____
   _____
   _____
   _____
   _____
   _____

2. How does the trend of population growth affect agricultural jobs?

   _____
   _____
   _____
   _____
   _____
   _____

3. What are the advantages of post-secondary education when seeking a career field?

   _____
   _____
   _____
   _____
   _____
   _____

4. What are the attributes employers look for when hiring an applicant?

   _____
   _____
   _____
   _____
   _____
   _____

Chapter 3
Safety in Livestock Production

INTRODUCTION

Agriculture is known to be one of the most dangerous occupations in the United States. Data show a death rate of 26 deaths per 100,000 workers. An overwhelming statistic shows that all industries combined have a death rate of 3.5 deaths per 100,000 workers. Many farm accidents result in nonfatal injuries that can be disabling and life changing. A majority of the nonfatal accidents involve some type of farm machinery, and the second most common agricultural accident involves handling livestock. Each year, approximately 150,000 farm accidents in the United States cause serious injury and disabling.

Farm accidents can be costly as well. The annual cost of farm accidents in the United States is between $4 and $5 billion. To avoid the financial and physical impacts of farm accidents, the law requires that agricultural employers maintain a safe work environment as enforced by the Occupational Safety and Health Act (OSHA). Chemicals are used for many applications in agriculture, and therefore, guidelines are in place to control their use. Policies that govern the use of chemicals in agriculture are created and controlled by the Environmental Protection Agency (EPA). A Materials Safety Data Sheet (MSDS) is required to be available for all chemicals in the workplace. The MSDS contains the details and characteristics of the chemical and any other hazardous information. Many injuries and even deaths related to improper chemical storage and labeling occur each year.

Given the dangerous nature of the agricultural industry, first aid preparation is necessary for all farming facilities. First-aid kits should be located in several key areas, such as the livestock buildings, on the farm equipment, and in all vehicles.

When handing any type of livestock, there is risk of injury to the workers and to the animals. Many accidents occur during loading and unloading of livestock onto trailers. Other hazards include fire, electrical current, manure pits, lagoons, livestock confinement buildings, and grain storage areas. Proper, heavy-duty facilities and equipment help reduce the risk of accidents.

Name_____ Date_____

Proper handling techniques and awareness from the workers will also reduce the number of injuries while handling livestock.

HAZARDS IN LIVESTOCK HANDLING

1. The most common types of animal handling injuries include the following:
   - Animal steps on handler.
   - Animal slips and falls on handler.
   - Animal pins or squeezes handler against a barrier.
   - Animal kicks handler.

2. By employing practical experience and adhering to a few general rules, handlers can prevent most accidents and injuries.
   - Move calmly, deliberately, and patiently. Avoid quick movements or loud noises that may startle animals.
   - Do not alter the daily routine or the animals' living conditions. Animals often balk at anything out of the ordinary.
   - Always leave an escape route when working in close quarters with animals.
   - Avoid startling an animal. Make it is aware of your approach before getting too close to it.

3. Understanding the blind spots of an animal is key to safely moving livestock. Figure 3-1 shows the area of sight for cattle. Avoiding the blind areas can help to reduce injuries.

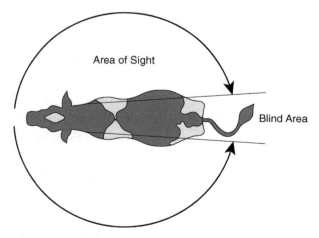

Figure 3-1 Area of sight for cattle.

Name_____ Date_____

EXERCISE 3-1 FACILITIES DESIGN WITH SAFETY IN MIND

Using the principles discussed, your task will be to design a safe confinement area to move cows on a local beef farm. The cattle are currently in a large, open pasture, making them hard to corral. The cows need to be corralled and individually restrained to allow the veterinarian to check them for pregnancy. If they are pregnant, they will be released into a brood cow pasture close to the barn; if they are not pregnant, they will be loaded onto a trailer for sale. The farm owner wants a permanent corral system that is ideally designed with cattle behavior in mind as well as handler and animal safety. A map of the current facilities is given here; it includes the current location of the cattle and the desired end location of the brood cow pasture and the trailer.

Your Task: Using the base map of the existing farm, design a safe corral system, including an effective funnel system to capture cattle from pasture, proper flow for cattle movement, equipment to restrain livestock while veterinarian checks for pregnancy, and proper facilities to separate cattle to pasture or to load on to the trailer. Include worker locations on the map as well

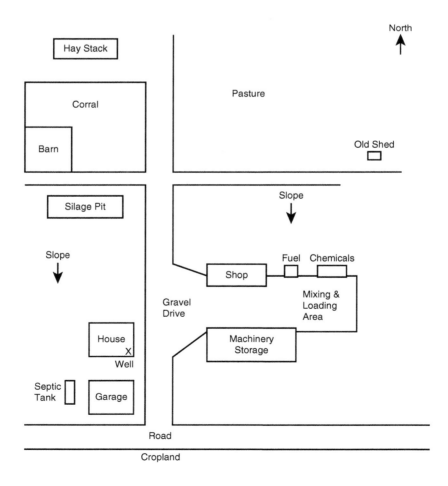

Name_____ Date_____

Explain: Describe your facilities and your reasons for designing it this way. Why did you organize the facility in this way? What animal behavior and safety principles did you use in your construction?

Name _____ Date _____

EXERCISE 3-2 LIVESTOCK HANDLING INJURY REPORT

Read the following scenario describing a common accident involving a livestock production facility. After reading the scenario, use the Supervisor's Incident Investigation Report to identify the causes of the accident and recommend changes to prevent similar incidents in the future.

Scenario

On The Chenery farm, a thoroughbred racehorse facility, the local emergency squad received a call for a multiple-victim farm accident. When they arrived at the scene, they found an adult male in his 40s and a 15-year-old teenage boy—both had been struck or trampled by two horses that were being loaded onto a trailer. The adult was in the trailer unconscious with lacerations and contusions to the head, as well as several broken bones. The teenager had injuries to the face and legs. The horses were loose and roaming about the farm when medics arrived. Both individuals were airlifted to the closest hospital for critical care.

Once the victims are stabilized and the horses captured and settled down, the investigation must begin. As the supervisor, you talk to the witnesses and assess the scene to determine what happened. Once the victims are stable, they can be surveyed for information about the accident. In this case, Charlie was a full-time employee who has worked as a groom and hot walker for five years. Charlie is in his 40s and has more than 20 years of experience with horses, but he is sometimes too comfortable with the horses. Andy is a teenager hired 2 weeks prior to the incident. Andy has very little experience with horses and does not understand animal behavior very well.

On this day, just around 5 a.m., Charlie was loading two thoroughbreds onto a trailer to head to Kentucky. They were running late and needed to get on the road. The loading ramp normally used for getting horse on the trailer was blocked by a tractor with a flat tire, so Charlie decided the quickest alternative was to load the horsed onto the trailer from the rear so they would not need the loading ramp.

Andy fetched the horses from their stalls; in a hurry, he slapped on their lead ropes, yelled at the horses to come, and pulled the lead hard as he moved fast toward the trailer. As he got to the trailer holding the leads to both horses, the smooth concrete was slick due to an overnight rain. Both horse lost their footing and stumbled just before stepping on the trailer. The horses reared up and panicked, refusing to get onto the trailer. Andy held onto the leads, but he was losing control, so Charlie took the leads from Andy and tried again to lead the horses onto the trailer. Charlie got both horses right up to the door of the trailer, but they were spooked and refused to enter the trailer. Charlie was inside the trailer tugging on the leads. In an effort to move the horses, Andy raised his voice, clapped, and waved his hands from behind to move the horses forward. Both horses lunged forward onto the trailer, pinning Charlie in the trailer; the escape door had never been unlocked. The trailer floor was also slick, so one horse fell to the ground and knocked Charlie down as well; both horses started flailing and trampling Charlie on the trailer floor. Seeing Charlie was in trouble, Andy tried to enter from behind but was kicked in the face by one of the horses and trampled as the horses backed out of the trailer. Both horses ran off to get away from the commotion, and the two workers were left on the ground, injured and helpless. Another worker arrived as the incident happened; she witnessed the injuries, called 911 immediately, and began first aid on the victims.

© 2016 Cengage Learning®. May not be scanned, copied or duplicated, or posted to a publicly accessible website, in whole or in part.

Supervisor's Incident Investigation Report

Instructions: Complete this form to describe the incident that resulted in serious injury or illness. Use the scenario to gather information for the report.

This is a report of a:	❏ Death ❏ Lost Time ❏ Dr. Visit Only ❏ First Aid Only ❏ Near Miss
Date of incident:	This report is made by: ❏ Employee ❏ Supervisor ❏ Team ❏ Final Report

Step 1: Injured employee (complete this part for each injured employee)

Name:	Sex: ❏ Male ❏ Female	Age:
Department:	Job title at time of incident:	
Part of body affected: (shade all that apply)	Nature of injury: (most serious one) ❏ Abrasion, scrapes ❏ Amputation ❏ Broken bone ❏ Bruise ❏ Burn (heat) ❏ Burn (chemical) ❏ Concussion (to the head) ❏ Crushing Injury ❏ Cut, laceration, puncture ❏ Hernia ❏ Illness ❏ Sprain, strain ❏ Damage to a body system: _____ ❏ Other _____	This employee works: ❏ Regular full time ❏ Regular part time ❏ Seasonal ❏ Temporary Months with this employer: Months doing this job:

Name_____ Date_____

Step 2: Describe the incident	
Exact location of the incident:	**Exact time:**
What part of employee's workday? ❑ Entering or leaving work ❑ Doing normal work activities ❑ During meal period ❑ During break ❑ Working overtime ❑ Other	

Names of witnesses (if any):

Number of attachments:	Written witness statements:	Photographs:	Maps/drawings:

What personal protective equipment was being used (if any)?

Describe, step-by-step, the events that led up to the injury. Include names of any machines, parts, objects, tools, materials, and other important details.

Description continued on attached sheets: ❑

Name_____ Date_____

Step 3: Why did the incident happen?	
Unsafe workplace conditions: (Check all that apply) ❏ Inadequate guard ❏ Unguarded hazard ❏ Safety device is defective ❏ Tool or equipment defective ❏ Workstation layout is hazardous ❏ Unsafe lighting ❏ Unsafe ventilation ❏ Lack of needed personal protective equipment ❏ Lack of appropriate equipment/tools ❏ Unsafe clothing ❏ No training or insufficient training ❏ Other: _____	Unsafe acts by people: (Check all that apply) ❏ Operating without permission ❏ Operating at unsafe speed ❏ Servicing equipment that has power to it. ❏ Making a safety device inoperative ❏ Using defective equipment ❏ Using equipment in an unapproved way ❏ Unsafe lifting by hand ❏ Taking an unsafe position or posture ❏ Distraction, teasing, horseplay ❏ Failure to wear personal protective equipment ❏ Failure to use the available equipment/tools ❏ Other: _____

Why did the unsafe conditions exist?

Why did the unsafe acts occur?

Is there a reward (such as "the job can be done more quickly" or "the product is less likely to be damaged") that may have encouraged the unsafe conditions or acts?   ❏ Yes   ❏ No

If yes, describe:

Were the unsafe acts or conditions reported prior to the incident?   ❏ Yes   ❏ No

Have there been similar incidents or near misses prior to this one?   ❏ Yes   ❏ No

Name_____ Date_____

Step 4: How can future incidents be prevented?

What changes do you suggest to prevent this injury/near miss from happening again?

❏ Stop this activity ❏ Guard the hazard ❏ Train the employee(s) ❏ Train the supervisor(s)

❏ Redesign task steps ❏ Redesign workstation ❏ Write a new policy/rule ❏ Enforce existing policy

❏ Routinely inspect for the hazard ❏ Provide personal protective equipment

❏ Other: _____

What should be (or has been) done to carry out the suggestion(s) checked above?

Description continued on attached sheets: ❏

Step 5: Who completed and reviewed this form? (Please Print)

Written by:	Title:
Department:	Date:

Names of investigation team members:	

Reviewed by:	Title:
	Date:

Name_____ Date_____

CHAPTER 3 LAB QUESTIONS

1. List five types of agricultural accidents.

   _____
   _____
   _____
   _____
   _____
   _____

2. Create a list of essential items that should be included in a first-aid kit.

   _____
   _____
   _____
   _____
   _____
   _____

3. What are five practices that can keep workers safe when handling livestock?

   _____
   _____
   _____
   _____

4. Why is it recommended to use curved chutes with solid colored panels for moving livestock?

   _____
   _____
   _____
   _____
   _____

Chapter 4
Livestock and the Environment

INTRODUCTION

Wastes associated with livestock production, have become more of a public concern than in the past. About 2 billion tons of manure are produced each year on livestock farms in the United States. Due to urban sprawl, attention has been drawn to the methods of animal waste removal and the impacts the waste disposal has on the environment and the local community. Animal wastes can include manures, soiled bedding products, juices associated with silage feeds, wash-down water from cleaning livestock facilities, manure runoff liquids, disposed animal carcasses, and others types of wastes. To the livestock producer, wastes can be assets if managed properly and can reduce the need for commercial fertilizers that may be harmful to the environment. Wastes also add organic material to the soil improving the fertility of the land.

If not managed properly, animal wastes can have a significantly negative effect on the environment. Our natural resources such as water, soil, and air may be jeopardized if animal wastes are not managed professionally. Protecting ground and surface water from waste contamination is a priority. Water is usually contaminated by leaching of waste products through the ground or by wastes running along the surface of the ground into a water source. The worst offence of water contamination is direct discharge into lakes and streams.

In livestock husbandry, waste accumulates every day. In order to manage these wastes properly to protect the environment and surrounding community, a livestock producer must develop a waste handling plan. A waste handling plan should include:

1. Information from area soil surveys.
2. A current nutrient analysis for all areas on which waste will be used, including anticipated crops to be grown.
3. A map of the farm showing roads, wells, waterways and basins, dwellings, and waste treatment and holding facilities.
4. Types and amounts of wastes and their nutrient values.
5. Rates and times of application.

In this chapter, you will learn to understand the environmental laws affecting the management of animal wastes. You will determine what the specific laws are in your state and how these environmental laws affect an animal producer. Next, you will calculate the amount of wastes that are produced by different types of livestock and calculate the amount of manure needed to fertilize a given amount of field area. You will also create a map of an example farm, keeping waste handling considerations in mind.

Manure being spread onto crop fields as fertilizer.

Name _____ Date _____

EXERCISE 4-1 ENVIRONMENTAL IMPACT OF WASTES (FEDERAL AND STATE LAWS)

Directions: For each of the following environmental laws, answer these questions: (1) How does the law protect the environment? (2) How does this limit or affect farming practices? (3) Which government agency enforces this law?

The Federal Water Quality Act of 1965

1. _____
2. _____
3. _____

The Refuse Act of 1899

1. _____
2. _____
3. _____

The Solid Waste Disposal Act of 1965

1. _____
2. _____
3. _____

The Federal Clean Air Act

1. _____
2. _____
3. _____

Your State Laws

1. _____
2. _____
3. _____

34 SECTION 1 The Livestock Industry

Name_____ Date_____

EXERCISE 4-2 MANURE APPLICATION CALCULATIONS

When applying wastes to the land, the farmer must know how much waste is produced each day, as well how much to apply to various areas of crop land.

Directions: Use the following scenario and values to answer the questions about waste application rates. Use Tables 4-2 and 4-3 in the textbook to find the values needed in the calculations.

Scenario

Joe is the crop specialist for a large dairy operation. One of his responsibilities is to manage the waste handling operation at the farm. As the manure is gathered at the facilities, Joe has a team of operators spreading the wastes onto their crop fields to fertilize the land. In this dairy operation, they milk 1,200 dairy cows every day and have to remove all of the waste from the barns. Joe has an expected corn yield of 170 bushels per acre. The livestock waste to be applied comes from pit storage that Joe uses to contain manure when conditions are not favorable for spreading it on fields. Joe has tested the nitrogen content of the manure and found it to be 40 pounds of nitrogen per 1,000 gallons of manure. When the time is right, Joe will apply the manure on the fields. He has 700 acres of land to fertilize with this manure.

Calculation Questions

1. How many pounds of wastes are produced at Joe's farm each day?
   _____
   _____

2. How many pounds of wastes are produced over the entire year (365 days)?
   _____
   _____

3. How many pounds of nitrogen are needed per acre?
   _____
   _____

4. How many gallons of waste will need to be applied per acre to grow the corn?
   _____
   _____

5. How many total gallons of waste will Joe need to apply to cover all of the farm property?
   _____
   _____

Name_____ Date_____

6. Will Joe's farm produce enough waste to properly fertilize the total farm property?
_____
_____

7. If Joe doesn't produce enough waste to fertilize his own farm, what are some alternatives?
_____
_____

CHAPTER 4 MATCHING

Term **Definition**

___ 1. point source
___ 2. agronomic nitrogen rate
___ 3. diversion
___ 4. drainage channel
___ 5. debris basin
___ 6. holding pond
___ 7. disposal
___ 8. masking agent
___ 9. digestive deodorant
___ 10. effective ambient temperature (EAT)

a. collects runoff water from a feedlot.

b. final step in controlling runoff from feedlots; the collected water can be used for irrigation of the land

c. bacteria that create a digestive process that eliminates the odor

d. temporary storage area for runoff; it is not designed for waste treatment.

e. cover up the odor of wastes with the introduction of another odor

f. a large source of pollution; the law prohibits discharge of pollutants from a point source into a river or stream without a permit

g. the combined effect of factors such as humidity, precipitation, wind, and heat radiation on the efficiency of energy use by farm animals

h. preventing surface water from outside the feedlot from getting onto the feedlot

i. keeps about 50 to 85 percent of the solids from reaching the holding ponds

j. the amount of available nitrogen per unit of yield necessary to produce a given crop

Name_____ Date_____

CHAPTER 4 LAB QUESTIONS

1. How much manure is produced on livestock operations in the United States every year?

   _____
   _____
   _____
   _____
   _____
   _____

2. What are the various types of manure?

   _____
   _____
   _____
   _____
   _____
   _____

3. Why is this manure an environmental concern? What are the possible environmental impacts?

   _____
   _____
   _____
   _____
   _____
   _____

4. Develop a waste handling plan for a local livestock producer.

   _____
   _____
   _____
   _____
   _____
   _____

SECTION 2

Anatomy, Physiology, Feeding, and Nutrition

Chapter 5	Anatomy, Physiology, and Absorption of Nutrients	40
Chapter 6	Feed Nutrients	51
Chapter 7	Feed Additive and Growth Promotants	61
Chapter 8	Balancing Rations	70

Chapter 5

Anatomy, Physiology, and Absorption of Nutrients

INTRODUCTION

It is important that professionals in the animal industry have a working knowledge of animal anatomy and physiology. Such knowledge is essential when making management decisions and implementing health care treatment. Decisions affecting feeding strategies, shelter, reproductive programs, and much more depend on the knowledge of those managing the animals. For example, a farm hand at a livestock operation must be able to identify abnormalities, injuries, or even a change in behavior and/or appearance. If the farm employee can identify the problem, he or she can notify the farm owner and treatment can begin. The owner and veterinarian can be called in to give a diagnosis and prepare a treatment strategy. Even owners of companion animals should have a knowledge of basic animal anatomy in order to recognize health issues and get treatment.

Many different kinds of tissues make up the bodies of animals. Tissues are groups of specialized cells that form individual organs, as well as complete systems that carry out the functions of the body. The animal body can be classified into 11 major systems: skeletal, muscle, respiratory, circulatory, nervous, endocrine, excretory, reproductive, immune, integumentary, and digestive. The skeletal system gives structure, support, and protection to the organs and systems of the body. The muscular system allows locomotion of the body and functions to move the skeletal system. Smooth and striated muscles perform different tasks, from creating the power for a horse to gallop or functions to making the heart beat to inhaling and exhaling from the lungs. The circulatory system's main function is to pump blood from the heart throughout the whole body. The blood circulation provides the cells of the body with nutrients and oxygen in order to function. Circulation also removes wastes from those cells. The respiratory system also provides important benefits to the body. The lungs contract and expand to draw in air, during which the oxygen is extracted and molecules are oxidized in order to provide energy for various body cells; waste products such as carbon dioxide are expelled from the body.

The nervous system acts as a communication system for the body, transmitting signals from one body system to another in order to coordinate the functioning on the body. The endocrine system helps to maintain the function, growth, and development of the body by secreting various hormones for each body system. The excretory system serves the function of removing waste products from the body. The integumentary system protects the body from damage. The immune system uses antibodies to defend the body against infection and disease. The reproductive systems is a specialized group of organs specific to each gender that allow for reproduction within animal species. The digestive system breaks down feeds into simple substances that can be absorbed into the bloodstream and used by body cells. Livestock animals can be categorized into two types of digestive systems: ruminants and non-ruminants. Ruminants, such as cattle, have a stomach with four compartments, which allows for the breakdown of more roughages by the beneficial bacteria in the rumen. Non-ruminants, such as horses, have a simple, one-compartment stomach. The simple stomach requires more concentrated feeds such as grain so the feed is easier to break down.

This chapter will explore each of the body systems through an anatomy model activity as well as a research activity involving animal hormones. Chapter matching and vocabulary will help to develop an understanding of anatomy terms.

Name_____ Date_____

EXERCISE 5-1 ANATOMY MODEL LAB

The purpose of this exercise is to assemble and label a visible horse anatomy model (Figure 5-1). The instructor will provide materials for the anatomy model.

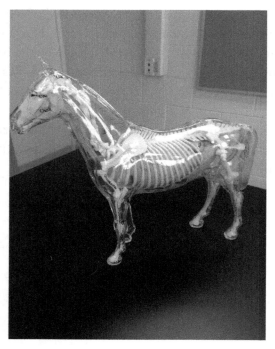

Figure 5-1 Example of visible horse model.

CHAPTER 5 Anatomy, Physiology, and Absorption of Nutrients   43

Name_____ Date_____

Visible horse model

Directions:

1. For the following diagram of the parts of the horse (Figure 5-2), fill in the blank space with the correct labels identifying each anatomical part. A similar diagram can be located in Chapter 31 of the textbook.

Figure 5-2   External horse parts: Fill in the blank.

44 SECTION 2 Anatomy, Physiology, Feeding, and Nutrition

Name_____ Date_____

2. The diagrams in Figures 5-3 and 5-4 have the labeled anatomy of the skeletal system as well as the basic internal anatomy. Using these diagrams, construct a visible horse model. All parts of the horse must be labeled; using a black marker will work best. The instructor will provide a manufactured model kit or will provide materials for constructing the skeletal system and the internal organs, as well as the external parts of the anatomy. When the model is complete and labeled, the following three criteria will be used to grade the model:

- Accuracy and location of all internal and external parts of the anatomy.
- Construction quality (clean, neat, glue accuracy, etc.).
- Labeling (accurate and legible).

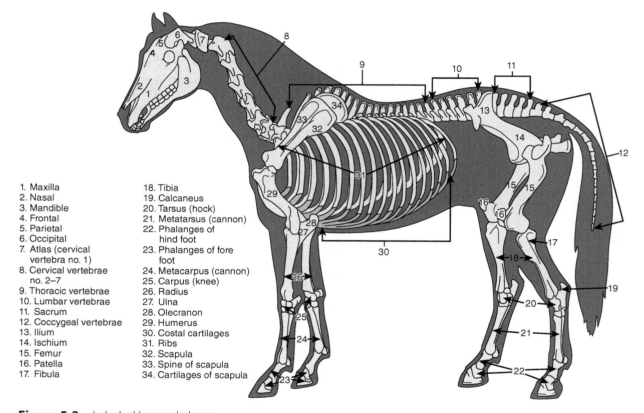

1. Maxilla
2. Nasal
3. Mandible
4. Frontal
5. Parietal
6. Occipital
7. Atlas (cervical vertebra no. 1)
8. Cervical vertebrae no. 2–7
9. Thoracic vertebrae
10. Lumbar vertebrae
11. Sacrum
12. Coccygeal vertebrae
13. Ilium
14. Ischium
15. Femur
16. Patella
17. Fibula
18. Tibia
19. Calcaneus
20. Tarsus (hock)
21. Metatarsus (cannon)
22. Phalanges of hind foot
23. Phalanges of fore foot
24. Metacarpus (cannon)
25. Carpus (knee)
26. Radius
27. Ulna
28. Olecranon
29. Humerus
30. Costal cartilages
31. Ribs
32. Scapula
33. Spine of scapula
34. Cartilages of scapula

Figure 5-3 Labeled horse skeleton.

Name_____ Date_____

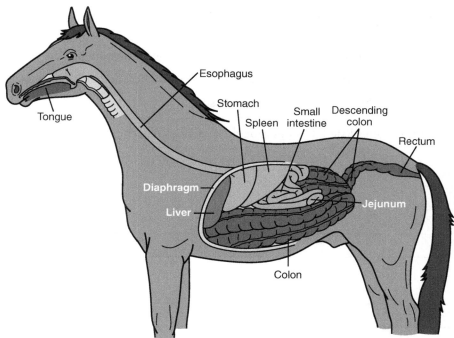

Figure 5-4 Internal anatomy of a horse.

Extension Activity

Paint a skeleton on the coat of a well-behaved horse. Use paint that is suitable for application on animals and easy to wash out. Perform a demonstration to a young group of students that teaches the shapes and names of the parts of the skeletal system. When finished, wash the paint off the horse.

Name_____ Date_____

EXERCISE 5-2 FUNCTIONS OF HORMONES

The endocrine system is reliant on hormones to act as chemical messengers to maintain the growth and development of the body. The glands of the endocrine system produce specific hormones, and as blood passes through, the gland absorbs the hormone and transports it to its target organ or tissue.

Directions: For each of the hormones listed, (1) describe the function of each hormone and (2) identify which gland produces the hormone.

1. Aldosterone
   a.

   b.

2. Epinephrine, also called adrenaline
   a.

   b.

3. Luteinizing hormone-releasing hormone (LHRH)
   a.

   b.

4. Estrogen
   a.

   b.

Name_____ Date_____

5. Progesterone

   a. _____
   _____

   b. _____
   _____

6. Insulin

   a. _____
   _____

   b. _____
   _____

7. Follicle-stimulating hormone (FSH)

   a. _____
   _____

   b. _____
   _____

8. Gonadotropin

   a. _____
   _____

   b. _____
   _____

9. Luteinizing hormone (LH)

   a. _____
   _____

   b. _____
   _____

Name_____ Date_____

10. Somatotropin

   a.
   _____
   _____

   b.
   _____
   _____

11. Oxytocin

   a.
   _____
   _____

   b.
   _____
   _____

12. Testosterone

   a.
   _____
   _____

   b.
   _____
   _____

Name_____ Date_____

CHAPTER 5 MATCHING ACTIVITY

Term

____ 1. exoskeleton
____ 2. fibrous joints
____ 3. amphiarthroses joints
____ 4. cells
____ 5. diarthroses
____ 6. synovial membrane
____ 7. striated voluntary muscle
____ 8. unstriated muscle
____ 9. system
____ 10. cardiac muscle
____ 11. larynx
____ 12. trachea
____ 13. alveoli
____ 14. arterioles
____ 15. pulmonary circulation system
____ 16. ossein
____ 17. plasma
____ 18. rumination
____ 19. absorption
____ 20. osteocytes

Definition

a. how cells divide and grow and differentiate into tissues with a variety of functions

b. a group of organs that carries out a major function

c. a gelatin-like protein found in bones

d. specialized cells that break down the cartilage and replace it with bone

e. typically, a hard shell on the outside of the body

f. found only in the muscular wall (myocardium) of the heart; it is striated in the same manner as skeletal muscle

g. flesh that has dark bands that cross each muscle fiber

h. where the actual exchange of gases occurs; the walls are very thin, about one cell thick, and are covered by a film of fluid that acts as a surfactant

i. carries blood through the lungs of the animal

j. connects to the capillary bed surrounding the alveoli

k. joined by fibrous tissue or, in some cases, cartilage tissue; these joints do not permit any type of movement

l. a tube that leads to the bronchi; the wall is lined with a series of C-shaped rings of cartilage

m. consist of discs of a fibrous cartilage that separate and cushion the vertebrae, allowing very limited movement

n. an area composed of cartilage structures; it contains vocal cords that vibrate when air passes across them, producing sound

o. refers to taking the digested parts of the feed into the bloodstream

p. joints that allow free movement and have a fluid-filled cavity

q. when ball-like masses are formed in the stomach and the material is then forced back up the esophagus to be chewed again

r. muscles that surround the hollow internal organs of the body, such as the blood vessels, stomach, intestines, and bladder

s. the fluid portion of whole blood

t. usually found inside the ligaments that help hold the joint together

Name_____ Date_____

CHAPTER 5 LAB QUESTIONS

1. What are the main functions of the skeletal system?

2. Name and describe the three types of muscle normally found in animal bodies.

3. How does the avian respiratory system differ from the mammalian system?

4. What is the importance of the endocrine system?

5. Why can ruminants digest large amounts of roughage?

Chapter 6
Feed Nutrients

INTRODUCTION

Domesticated animals rely on humans to provide feed that contain adequate nutrients to sustain various stages of growth and development. A nutrient is a chemical that aids in the support of life and provides energy by being absorbed into the body cells. Nutrients are divided into five groups: (1) energy nutrients (carbohydrates, fats, and oils), (2) proteins, (3) vitamins, (4) minerals, and (5) water. Animals require many different nutrients to maintain the daily functions of the body. Each nutrient needs to be provided to the animal in balance with the other nutrients. To provide proper nutrition, an animal producer must create a feed that has balanced rations of each nutrient. These rations are calculated using percentages, and they can be adjusted to provide different levels of nutrients, depending on the animals' stage of growth.

Animals differ in the kinds and amounts of nutrients they require. A lactating dairy cow has a higher nutrient requirement than a retired racehorse. It is the job of animal nutritionists to determine which nutrients animals need and then make recommendations for feeding each kind of animal. The recommendations are based on the results of feeding experiments and nutrient sampling and testing, as well as nutrient content calculations. A good nutritionist will recommend a feeding strategy that provides proper nutrition for growth and production, while also being economical. Too much of one nutrient and not enough of another may result in unhealthy animals, low production, and high feed costs.

Small-scale animal producers or hobby farmers may purchase bagged feed from a grain elevator or a local feed store. When purchasing bagged feed, a producer should understand the information on a feed tag in order to purchase the correct feed for his or her animal. On the tag, the maximum amount of available fiber in that feed is shown as a percentage of the total weight of the bag of feed. A list of ingredients will also be printed on the tag to show the sources of carbohydrates contained. Some common examples of carbohydrates in bagged feed are cane molasses, ground corn, wheat middlings, and oats.

In this chapter, you will conduct a feed sampling activity as well as map the path that grain travels from the fields in which produced to the grain elevator and eventually to the farmer.

Name_____ Date_____

EXERCISE 6-1 FEED SAMPLING ACTIVITY

Accurate feed analyses are necessary for animal producers to balance feeding rations, accurately price hay, formulate needed supplements, and determine the nutrient value of feeds made on the farm. Adequate forage testing is necessary to develop an accurate feed inventory. An inventory is necessary to allocate the high nutrient feeds to high producing livestock and the lower nutrient feed to animals at a lower level of nutrient requirements.

Directions: Using the sampling guidelines page, sample various "lots" of feed. Access to feed inventories will be made available by the instructor. Accurately obtain the samples and prepare them for testing. For each type of feed, record the date and time, identify the "lot" of feed, and describe how the sample was prepared and handled in detail.

Feed Sampling Report

Test 1: Round bales

Date and time: _____

Lot of feed: _____

Sample preparation/handling details:
_____
_____

Test 2: Compressed loaf hay stack

Date and time: _____

Lot of feed: _____

Sample preparation/handling details:
_____
_____

Test 3: Chopped or ground hay

Date and time: _____

Lot of feed: _____

Sample preparation/handling details:
_____
_____

Test 4: Harvest silage (fresh as it comes out of the field)

Date and time: _____

Lot of feed: _____

Sample preparation/handling details:
_____
_____

Name_____ Date_____

Test 5: Silage at feeding from an upright silo

Date and time: _____

Lot of feed: _____

Sample preparation/handling details:
_____
_____

Test 6: Silage from horizontal silo

Date and time: _____

Lot of feed: _____

Sample preparation/handling details:
_____
_____

Sampling Guidelines

The feed values of most forages vary due to factors such as time of harvest, soil quality, moisture content, and many other factors. To accurately manage a feed inventory, test forages routinely to determine when and how the feed should be used. An accurate forage inventory is necessary so the producer can allocate high-protein feeds to animals with a higher nutrient demand and allocate slightly poorer quality feeds to livestock that have a low nutrient requirement. This guideline focuses on the sampling strategies when testing hay and silage.

The Sample "Lot"

Samples must closely resemble the entire lot of forage. Each sample should only represent one lot. A *lot* is a term used to describe forage that was all harvested from the same field within a 48-hour period and usually contains less than 100 tons of forage. The most important characteristic of a lot is that it is uniform. All forage from the same lot should have following similarities: type of plant(s), field soil type, cutting date, maturity, variety, weed contamination, type of harvest equipment, weather during growth and harvest, preservatives, drying agents, additives, storage conditions, and pest or disease damage. The consistency of these factors is essential to obtaining an accurate sample. If any of these factors differ, designate a new lot of forage.

Hay Sampling

Sample baled long hay after curing. A core sampler or a probe is an essential tool for sampling hay. The core sampler should be inserted into the bale at least 12 to 18 inches and have and internal diameter of at least $\frac{3}{8}$ inch. Core samplers usually have sharpened or serrated tips to cut through hay. Several core samplers are designed to be operated with an electric hand drill. The sampler should be inserted at a slow speed to avoid the sample heating up and then being unfit for testing.

Name_____ Date_____

Square or Round Bales

Select 20 or more average bales from each lot. Collect one sample from each bale by coring the square bale straight in from the center of the end of the bale. Take samples from any broken hay bale piles or hay stacks that will be used to feed livestock in some way.

Loosed or Compressed Hay Stacks

When collecting sample, use a hay probe at least 24 inches long to collect 15 or more samples from each lot. When sampling loose hay stacks, make sure to take core samples from six general locations: (1) top front, (2) top middle, (3) top rear, (4) lower front side, (5) lower middle side, and (6) lower rear side. When top locations are sampled, stand on the top of the stack and insert the probe vertically downward. For side sampling, use a light angle downward with the probe to avoid sampling parallel stems in the stack.

Chopped or Ground Hay

On a regular basis, collect 10 samples from each lot of hay during grinding, and place the samples in a sealed plastic bag. When sampling previously ground hay in a pile, collect about one-quarter of the samples from the top half of the pile and the rest from the lower half.

Silage Sampling

Silage can be sampled when entering the silo, when feeding after storage, or a combination of both. When forage is properly ensiled, results from fresh samples will agree closely with fermented forage. Avoid collecting samples from rotted or poorly preserved materials in the silage; this kind of spoilage is usually found at the top of upright silos or the shallow end and slopes of a bunk silo.

Fresh Samples

Collect 20 or more samples from each lot of fresh silage as it comes out of the field. Take periodic samples as loads are brought to silo while making sure the container remains sealed to reduce loss of moisture content. For example, when sampling a large silo, collecting one quart from each load in the morning, at noon, and in the afternoon will give an accurate sample of the hay inventory. The following procedure will help ensure an accurate sampling method:

1. Collect at least 2 gallons of samples and mix well.
2. Make a cone-shaped pile of the forage sample.

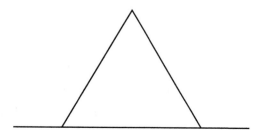

Name_____ Date_____

3. Divide the pile as if dividing a pie into four pieces.

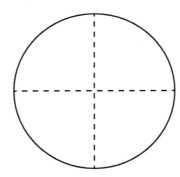

4. Randomly choose one of the quarters to keep and then choose the diagonally positioned quarter; discard the other two quarters.

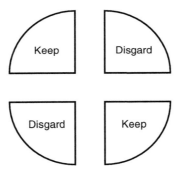

5. Repeat this procedure until 1 to $1\frac{1}{2}$ quarts remain. Then transfer all materials into a 2-quart plastic freezer bag. Remove the air and seal tight.

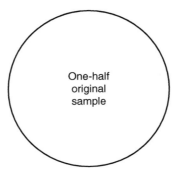

Name_____ Date_____

Upright Silos at Feeding

Collect 2 to 3 gallons of feed samples in 2-quart increments by passing a clean container under the chute while unloading ensilaged materials. Another method is to collect 20 handfuls of silage as it is fed out to livestock in different locations of the feed bunk. Mix and subsample, as described in the preceding five sampling steps.

Horizontal Silos

Collect 20 or more hand samples from various locations of the exposed face of the silo to represent the entire exposed surface. Mix and subsample as described earlier.

Name_____ Date_____

EXERCISE 6-2 GRAIN ELEVATOR MAP

Grain elevators are a key component of the current livestock industry in the United States, and the elevator manager must ensure that all of the feed company's clients get a high-quality feed with no mold or other contaminants. In this activity, assume the role of the grain elevator manager. The feed company has decided to expand the elevator to accommodate all of its clients' needs. The elevator will need to be equipped with enough silos and augers to provide all of the needed feed. The other responsibility will be to design the layout of the elevator to allow each silo to unload onto rail road cars.

Directions: Using the sample grain elevator maps for reference (Figures 6-1 and 6-2), design a grain elevator operation that can accommodate the needs of the feed company's clients. All of the grain silos must be able to unload onto a railroad car. Use technical drawing paper, a straight edge, and a pencil to create the elevator map.

Client Needs

Feed	Processing Procedure	Purpose of Processing
Barley (whole)		
Cracked shelled corn	Shelled, cracked (dry rolled)	Increase digestibility
Crimped oats	Crimped (steam rolled)	Increase digestibility
Dried beet pulp	Dried by-product of sugar production	Ease of handling
Ground shelled corn	Shelled, ground (dry rolled)	Increase digestibility
Liquid molasses	By-product of sugar production	
Shelled (whole kernel) corn	Shelled (removed from cob)	Increase nutrient concentration
Soybean meal	By-product of oil extraction	
Steam rolled oats	Steam rolled	Increase digestibility
Wheat middlings	By-product of flour milling	

Figure 6-1 Grain elevators.

58 SECTION 2 Anatomy, Physiology, Feeding, and Nutrition

Name_____ Date_____

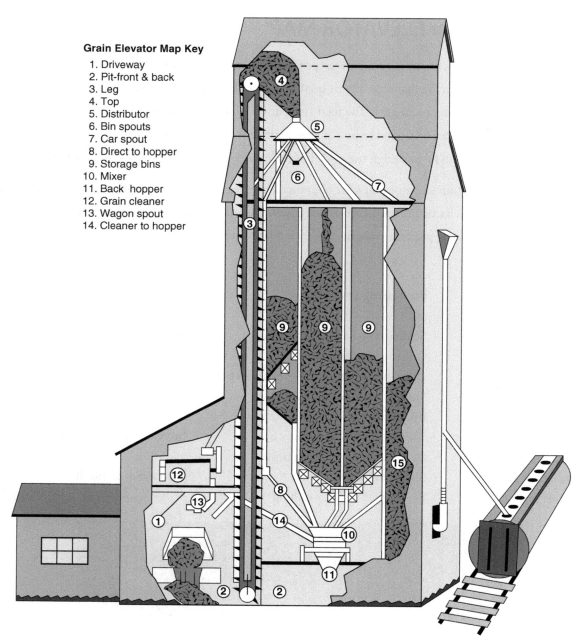

Figure 6-2 Map of parts of a grain elevator.

Name_____   Date_____

CHAPTER 6 — MATCHING

Term

____ 1. nutrient
____ 2. carbohydrate
____ 3. fiber
____ 4. commercial feed tag
____ 5. fat
____ 6. oil
____ 7. protein
____ 8. crude protein
____ 9. digestible protein
____ 10. deficiency

Definition

a. a solid at body temperature; may contain a higher energy value than carbohydrates

b. label attached to a bag of feed purchased at a grain elevator or feed store

c. organic compounds made up of amino acids

d. lack of a certain mineral in an animal's ration

e. The amount of ammoniacal nitrogen in the feed multiplied by 6.25

f. a chemical element or compound that aids in the support of life

g. approximately the amount of true protein in a feed

h. substances that are more difficult to digest than simple carbohydrates

i. the main energy nutrients found in animal rations; made up of sugars, starches, cellulose, and lignin

j. liquid at body temperature

CHAPTER 6 LAB QUESTIONS

1. What is a nutrient?

2. List the six groups of nutrients.

3. What is a balanced ration?

4. What factors will make feed rations differ from each other?

Chapter 7

Feed Additive and Growth Promotants

INTRODUCTION

Feed additives are materials used in an animal feed ration that are not classified as nutrients. These materials are used in small amounts to improve feed efficiency, promote faster growth gains, improve animal health, or increase production of animal products. Adding antibiotics at low levels in the ration over a period of time is a common practice in livestock feeding. Many different additives are available, and each has a different desired effect on livestock performance. Too much of one type of additive can cause negative health effects.

Antibiotics compounds are used to kill or slow the growth of some kinds of microorganisms. Other additives such as dewormers (anthelmintics) can be used to control various species of worms that can infest animals. Each species of animal has different needs and not all livestock are approved for every additive. Beef producers commonly use hormone implants; a high percentage of beef cattle rations use either monensin sodium (Rumensin) or lasalocid sodium (Bovatec) to increase the daily rate of weight gain. Feed additives are not as common in dairy cattle because they cannot have milk contaminated by hormones or antibiotics. Young dairy cattle that are not producing milk commonly receive a medicated feed until they near the age of lactation. There are very few additives that have been approved for mixture in sheep and goat feed rations. In swine feed, additives are a common practice to reduce mortality, especially in young stock. Some poultry producers also use medicated feeds to improve growth rates in broilers and egg production in layers. In horses, the use of feed additives is not common, except for occasional use in young colts and fillies.

The Food and Drug Administration (FDA) regulates the use of feed additives, including hormones and antibiotics. There has been heightened concern about the long-term health hazards associated with the use of medicated feeds and the ability of microorganisms to develop a resistance to the antibiotics and lose the ability to fight common livestock sicknesses. This chapter explores additives through a feed additives analysis and an antibiotic residue exercise.

EXERCISE 7-1 FEED ADDITIVE ANALYSIS

Directions: For the following descriptions of feed additives and their functions, (1) identify each additive, (2) identify which classification the additive it belongs in, and (3) describe how and why the additive has such an effect on the livestock.

Description 1: A beef farmer uses Bovatec trying to increase his cattle's rate of gain.

1. _____
2. _____
3. _____

Description 2: MGA is added to a heifer feed for a growing group of dairy heifers.

1. _____
2. _____
3. _____

Description 3: Implants are used on feeder cattle to increase the proportion of lean meat deposition on the carcass.

1. _____
2. _____
3. _____

Description 4: Monesin is used to improve feed efficiency and rate of gain in replacement dairy heifers.

1. _____
2. _____
3. _____

Description 5: Anhydrous ammonia (NH_3) is being used as a forage preservative for poor-quality forage.

1. _____
2. _____
3. _____

Name_____ Date_____

Description 6: Sodium bicarbonate (backing soda) is used on a group of dairy cattle for 4 months after calving.

1. _____
2. _____
3. _____

Description 7: Lactating dairy cows receive isoacids for 12 weeks while in high milk production.

1. _____
2. _____
3. _____

Name_____ Date_____

EXERCISE 7-2 ANTIBIOTIC RESIDUES

Feed additives containing medications can be an economical method for keeping animals healthy during growth. Proper management of medicated feeds is important to avoid sending animals to the butcher that may still have antibiotic residues in their blood stream. When caught by U.S. Department of Agriculture (USDA) testing, antibiotic residues will cause livestock to be rejected from slaughtering facilities. In this activity, answer questions based on a pig operation raising suckling pigs. The pigs are farrowed on April 2 and will be processed at a USDA inspected facility on May 21.

Included below is a sample Medicated Feed Label for swine.

Sample Medicated Feed Label

SuperStart

Medicated

For control of porcine colibacillosis (weaning pig scours) caused by susceptible strains of *ESCHERICHIA COLI*. Follow carefully the feeding directions and WARNING statement printed on the back of this label.

Active Drug Ingredient: Apramycin (as apramycin sulfate)...................150 grams per ton

GUARANTEED ANALYSIS

Crude Protein, not less than 21%
Crude Fat, not less than 10%
Crude Fiber, not more than 2.50%
Vitamin A, I. Units per lb. (min) 6,000
Vitamin D3, I. Units per lb. (min) 750.0
Vitamin E, I. Units per lb. (min) 55.0
Riboflavin, mgs. per lb. (min) 4.5
Niacin, mg per lb. (min) 30.0
d-Pantothenic Acid, mg per lb. (min) 15.0

Choline, mg per lb. (min) 550.0
Vitamin B12, mg per lb. (min) 0.022
Menadione (Vitamin K), mg per lb. (min) 4.5
Biotin, mg per lb. (min) 0.09
Folic Acid, mg per lb. (min) 0.09
Pyridoxine, mg per lb. (min) 0.018
Thiamine, mg per lb. (min) 0.009
Lysine, not less than 1.60%

INGREDIENTS: Dried skim milk, dried whey, animal plasma, heat processed soybeans, fish meal, feeding oat meal, ground corn, meat and bone meal, corn distillers dried solubles, natural and artificial flavors added, sugar, yucca schidigera extract, dehydrated yeast culture, animal fat, cane molasses, monosodium glutamate, methionine, lysine, vitamin A acetate, D-activated animal sterol (source of vitamin D3), riboflavin supplement, niacin supplement, calcium panthothenate, choline chloride, vitamin B12 supplement, menadione dimethylpyrimidionol bisulfite (source of vitamin K), dl alpha tocopheryl acetate (source of vitamin E), biotin, folic acid, pyridoxine hydrochloride, thiamine mononitrate, calcium carbonate, salt, dicalcium phosphate, magnesium oxide, manganous oxide, ferrous sulphate, copper sulfate, cobalt carbonate, ethylenediamine dihydriodide, zinc sulphate and sodium selenite. **SuperGrow Feed Co. • Toledo, Iowa 52342**

Name_____ Date_____

Feeding Directions: SuperStart AP-150, medicated is a highly palatable product formulated especially for baby pigs being weaned at three weeks of age or earlier and a special formulation for "tail enders" needing a nutritional boost.

Begin feeding SuperStart AP-150, medicated, when pigs are approximately 5 lbs. body weight (1 week of age) and feed continuously and as the sole ration until the pigs have consumed at least 5 lbs. per pig or at least 7 to 10 days after weaning. NEVER WEAN AND CHANGE FEED SOURCE AT THE SAME TIME.

For "tail enders", separate pigs according to size. It is recommended to group pigs by size and place them in groups of 20 or less with a weight difference of no more than 10%.

Feed SuperStart AP-150, medicated, continuously and as the sole ration for at least 3 to 4 weeks or until pigs regain their healthy bloom. SuperStart can also be used as a high nutrient dense product in any starting program to encourage early dry diet consumption.

NOTE: Strains of organisms vary in their degree of susceptibility to antibiotics. If no improvement is observed after the recommended treatment, diagnosis and susceptibility should be reconfirmed.
IMPORTANT: Store in a clean, dry area, free of all offensive odors.
WARNING:
DISCONTINUE USE OF THIS MEDICATED FEED 28 DAYS BEFORE SWINE ARE MARKETED FOR HUMAN CONSUMPTION

Directions: Study the feed label above and answer the following questions based on the information in the label:

1. What is the name of the feed?
   _____
   _____
   _____

2. Who manufactured it?
   _____
   _____
   _____

3. What size of animals should it be fed to?
   _____
   _____
   _____

4. What form is the feed?
   _____
   _____
   _____

Name _____ Date _____

5. What ingredients are in this feed?
   _____
   _____
   _____

6. What ingredient is in the largest quantity?
   _____
   _____
   _____

7. Is this a medicated feed?
   _____
   _____
   _____

   - What medication(s) is(are) in the feed?
     _____
     _____

   - What is the withdrawal time for animals on this medicated feed?
     _____
     _____

8. How much protein is in this feed?
   _____
   _____
   _____

9. How many pounds are in the final feed mix?
   _____
   _____
   _____

10. What can you tell by the order in which the ingredients are listed?
    _____
    _____
    _____

11. What are the feeding directions?
    _____
    _____
    _____

Name_____ Date_____

12. What cautions are listed?
   _____
   _____
   _____

13. For the suckling pigs in this example, when should the medicated feed be started? When should it be stopped?
   _____
   _____
   _____

Name_____ Date_____

CHAPTER 7 MATCHING ACTIVITY

Term

____ 1. feed additive
____ 2. growth promotant
____ 3. antibiotic
____ 4. antimicrobial
____ 5. subtherapeutic
____ 6. subclinical disease
____ 7. hormone
____ 8. anthelmintic
____ 9. melengestrol acetate (MGA)
____ 10. withdrawal period

Definition

a. the use of medicated feeds at a lower level to treat sick animals

b. the period of time necessary for an antibiotic's residues to be eliminated from the bloodstream

c. are present in the animal's body at levels too low to cause visible effects

d. are used to treat for worms in swine

e. materials used in animal rations to improve feed efficiency, promote faster gains, improve animal health, or increase production of animal products

f. suppresses estrus (prevents the heifer from coming into heat), which reduces the continual mounting seen when heifers are coming into heat in the feedlot

g. hormone implants

h. kill or slow down the growth of some kinds of microorganisms

i. refers to both natural hormones and synthetic hormone-like compounds

j. antimicrobial compound

Name_____ Date_____

CHAPTER 7 LAB QUESTIONS

1. What are the benefits of feed additives?

   _____
   _____
   _____
   _____
   _____
   _____

2. List examples of various additives used in livestock feeds.

   _____
   _____
   _____
   _____
   _____
   _____

3. In which livestock industry is the use of feed additives most common? Why?

   _____
   _____
   _____
   _____
   _____

4. Why is there concern over long-term use of antibiotic additives in feeds?

   _____
   _____
   _____
   _____
   _____
   _____

Chapter 8
Balancing Rations

INTRODUCTION

Providing an appropriate, balanced ration to livestock is the key to efficient production. The combination of concentrates, feed additives, and forages needs to be calculated to provide the most economic return for producers. Livestock feeds are classified as either roughage or a forage—roughages have a crude fiber content of more than 18 percent, and concentrates have crude fiber content of less than 18 percent. Roughages can be either legume or nonlegume crops. A legume crop such as soybeans uses nitrogen from the air and is higher in protein content than nonlegumes crops such as grass hay. Concentrates are made from either energy feeds such as corn and oats or from protein supplements provided by animal or vegetable sources.

The right balance of roughages and concentrates must be in a ration. A ration is the amount of feed an animal is given during a 24-hour period. A ration is only balanced if it provides an animal with all of the nutrients it needs for growth, daily rate of gain, or production. A properly balanced ration should have an acceptable taste for the livestock as well as be cost effective for the producer. Any feed additives must be managed to not give higher amounts of certain substances than may be healthy for the animal. When balancing a ration for livestock, the needs for protein, energy, minerals, and vitamins must all be considered. There are five common steps that should be followed as a general rule of thumb when balancing a ration:

1. The kind of animal to be fed is identified.
2. The needs of the animal are found.
3. Feeds are selected and the composition of the feed is found.
4. The amount of each feed to use is calculated.
5. The ration is checked against the needs of the animal to make sure it is balanced.

Modern livestock operations use computer programs developed to assist in balancing feed rations. Two examples of such computer programs are Nutrient Requirements of Beef Cattle and Nutrient Requirements of Swine. When balancing a ration by hand, the Pearson's square is a valuable tool. In this chapter, a dry matter calculation problem will be presented as well as a sample ration that must be balanced.

72 SECTION 2 Anatomy, Physiology, Feeding, and Nutrition

Name_____ Date_____

EXERCISE 8-1 DRY-MATTER CALCULATION

Dry matter (DM) is the moisture-free content of a feed sample. The moisture dilutes the concentration of nutrients but does not have a major influence on intake. It is important to always balance and evaluate rations on a dry-matter basis. The calculation for converting nutrient values into a percentage of dry-matter value is a as follows: The DM composition can be found by dividing the as-is value by the moisture content. (DM = As-is/Moisture content) For example, A feed with a 16.8 percent crude protein (CP) as-is ÷ 83.2 moisture content = 20.19% CP on a DM basis.

Directions: For each of the following feed analysis reports, convert each value into a dry-matter basis using the DM calculation.

CLIENT SAMPLE ID 1: Alfalfa Haylage

−ANALYSIS−

	AS-RECEIVED BASIS	DRY-MATTER BASIS
MOISTURE, %	67.6	0.0
DRY MATTER, %	32.4	100.0
CRUDE PROTEIN, %	5.9	_____
HEAT DAM. PROTEIN, %	0.5	_____
AVAILABLE PROTEIN, %	5.9	_____
DIG. PROTEIN EST., %	3.8	_____
ACID DET. FIBER, %	12.1	_____
NEUT. DET. FIBER, %	17.6	_____
TDN EST., %	19.1	_____
ENE EST., THERMS/CWT	16.1	_____
NE/LACT, MCAL/LB	0.19	_____
NE/MAINT, MCAL/LB	0.19	_____
NE/GAIN, MCAL/LB	0.10	_____
PHOSPHORUS (P), %	0.11	_____
CALCIUM (CA), %	0.42	_____
POTASSIUM (K), %	0.90	_____
MAGNESIUM (MG), %	0.11	_____

Name_____ Date_____

CLIENT SAMPLE ID 2: Brown Midrib Sorghum Silage

	-ANALYSIS-	
	AS-RECEIVED BASIS	DRY-MATTER BASIS
	_____	_____
MOISTURE, %	75.9	0.0
DRY MATTER, %	24.1	100.0
CRUDE PROTEIN, %	2.6	_____
HEAT DAM. PROTEIN, %	0.3	_____
AVAILABLE PROTEIN, %	2.5	_____
DIG. PROTEIN EST., %	1.7	_____
ACID DET. FIBER, %	9.4	_____
NEUT. DET. FIBER, %	14.0	_____
TDN EST., %	15.5	_____
ENE EST., THERMS/CWT	13.1	_____
NE/LACT, MCAL/LB	0.16	_____
NE/MAINT, MCAL/LB	0.16	_____
NE/GAIN, MCAL/LB	0.09	_____
PHOSPHORUS (P), %	0.06	_____
CALCIUM (CA), %	0.12	_____
POTASSIUM (K), %	0.49	_____
MAGNESIUM (MG), %	0.04	_____

Name_____ Date_____

CLIENT SAMPLE ID 3: Baled Corn Stalks

	−ANALYSIS−	
	AS-RECEIVED BASIS	DRY-MATTER BASIS
MOISTURE, %	14.7	0.0
DRY MATTER, %	85.3	100.0
CRUDE PROTEIN, %	5.3	_____
HEAT DAM. PROTEIN, %	0.9	_____
AVAILABLE PROTEIN, %	4.7	_____
DIG. PROTEIN EST., %	3.6	_____
ACID DET. FIBER, %	41.6	_____
NEUT. DET. FIBER, %	59.4	_____
TDN EST., %	40.9	_____
ENE EST., THERMS/CWT	33.8	_____
NE/LACT, MCAL/LB	0.41	_____
NE/MAINT, MCAL/LB	0.35	_____
NE/GAIN, MCAL/LB	0.14	_____
PHOSPHORUS (P), %	0.18	_____
CALCIUM (CA), %	0.35	_____
POTASSIUM (K), %	1.68	_____
MAGNESIUM (MG), %	0.13	_____

Name_____ Date_____

EXERCISE 8-2 BALANCING A RATION USING EQUATIONS

The Pearson's square (Figure 8-1) ration formulation procedure is designed for balance simple rations. In order for the square to work, follow specific directions for its use. The nutrient contents of ingredients and nutrient requirements must be expressed on the same basis (that is, dry-matter or "as-fed""").

Several numbers are in and around the square, and one of the important numbers is the number that appears in the middle of the square. This number represents the nutritional requirement of an animal for a specific nutrient. It may be crude protein or TDN, amino acids, minerals, or vitamins.

In order to make the square work consistently, there are three very important considerations:

1. The value in the middle of the square must be between the two values that are used on the left side of the square. For example, the 14 percent crude protein requirement has to be an intermediate value between the soybean meal that has 45 percent crude protein or the corn that has 10 percent crude protein.

2. Disregard any negative numbers that are generated on the right side of the square. Be concerned only with the numerical differences between the nutrient requirement and the ingredient nutrient values.

3. Subtract the nutrient value from the nutritional requirement on the diagonal and arrive at a numerical value entitled parts. By summing those parts and dividing by the total, you can determine the percentage of the ration that each ingredient should represent in order to provide a specific nutrient level. Always subtract on the diagonal within the square in order to determine parts. Always double-check calculations to make sure that you do not have any mathematical errors. It also is very important to work on a uniform basis. Use a 100 percent dry-matter basis for nutrient composition of ingredients and requirements and then convert to an as-fed basis after the formulation is calculated.

Corn represents (31.0/35.00) × 100 of the ration, or 88.57 percent.

Soybean meal represents (4.0/35.00) × 100 of the ration, or 11.43 percent.

Check of the calculation:

88.57 lb corn × 10.0% CP	=	8.86
11.43 lb SBM × 45.0% CP	=	5.14
100.00 lb mixture contains	=	14.00 lb CP, or 14 percent

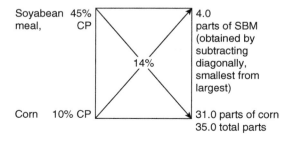

Figure 8-1 Pearson's square.

Name_____ Date_____

Directions: For the next two feeding scenarios, calculate the amount of each feed needed to create a balanced ration.

Scenario 1

You're feeding 20-pound baby pigs. Pigs this size need an 18 percent protein ration. You have shelled corn, which has 9 percent protein in it, but this is not enough protein. So, you decide to mix soybean oil meal with the shelled corn. The soybean oil has 40 percent protein in it. Mix the shelled corn and soybean oil meal together to come up with a 12 percent protein ration.

1. If you are going to mix 100 pounds of feed, how much corn do you need?
_____
_____

2. How much soybean meal do you need for 100 pounds of feed?
_____
_____

Scenario 2

Your pigs are now 50 pounds and you need to change the protein content in their ration. Now the ration must have 16 percent protein. You're going to mix corn, which has 9 percent protein, and soybean oil meal, which has 40 percent protein in it. What will you do?

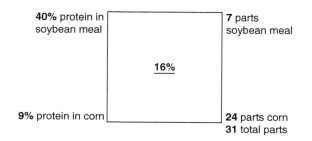

Name_____ Date_____

1. If you are going to mix a 100 pounds of feed, how much corn do you need?
_____
_____

2. How much soybean oil meal do you need?
_____
_____

3. Does this feed ration have more corn or more soybean meal than in the previous ration? Why?
_____
_____

CHAPTER 8 MATCHING ACTIVITY

Term

___ 1. legume
___ 2. protein supplement
___ 3. energy feed
___ 4. ration
___ 5. balanced ration
___ 6. micronutrients
___ 7. digestible energy
___ 8. metabolizable energy
___ 9. dry-matter basis
___ 10. as-fed (air dry) basis

Definition

a. livestock feeds with less than 20 percent crude protein
b. ingredients used in small quantities in the diet
c. the gross energy of a feed minus the energy remaining in the feces of the animal after the feed is digested
d. plants that have nodules (small swellings or lumps) on their roots that contain nitrogen-fixing bacteria
e. the mass of a material when the moisture content is removed
f. livestock feeds that contain 20 percent or more protein
g. data that are calculated on the basis of the average amount of moisture found in the feed as it is used on the farm
h. a ration that has all the nutrients the animal needs in the right proportions and amounts
i. the gross energy in the feed eaten minus the energy found in the feces, the energy in the gaseous products of digestion, and the energy in the urine
j. the amount of feed given to an animal to meet its needs during a 24-hour period

Name_____ Date_____

CHAPTER 8 LAB QUESTIONS

1. What is the difference between roughages and concentrates?

   _____
   _____
   _____
   _____
   _____
   _____

2. Describe the characteristics of a well-balanced ration.

   _____
   _____
   _____
   _____
   _____
   _____
   _____

3. What are five common steps to use when balancing a ration?

   _____
   _____
   _____
   _____
   _____
   _____

4. How does the Pearson's square work?

   _____
   _____
   _____
   _____
   _____
   _____

SECTION 3

Animal Breeding

Chapter 9	Genetics of Animal Breeding	82
Chapter 10	Animal Reproduction	90
Chapter 11	Biotechnology in Livestock Production	98
Chapter 12	Animal Breeding Systems	105

Chapter 9
Genetics of Animal Breeding

INTRODUCTION

Modern livestock have improved in quality from livestock of the past due to advancements and increased understanding of animal genetics. Livestock producers are able to use genetic technology to select for desirable traits among their herds, and over time, they can change the genetic makeup of the entire herd. Genetics is the study of heredity, or the way in which traits of parents are passed on to offspring. This concept was founded by an Austrian monk named Gregor Mendel. From 1857 to 1865, Mendel conducted many experiments with garden peas in order to prove that certain characteristics were passed on from the parent to the offspring.

The cells of an animal's body undergo cell division to allow the body to continually grow. Within the nucleus of each cell, DNA is contained in the form of chromosome pairs. One chromosome of the pair comes from the mother and the other comes from the father. The amount of difference between parents and offspring is caused by genetics and environmental factors. Heritability estimates are used to show how much of the difference in some traits may come from genetics.

Genes are complex molecules found on the chromosomes; genes have the ability to affect the dominant traits displayed in the animal. Some genes are dominant and others are recessive. Dominant genes overpower the effect of recessive genes. Some genes are neither dominant nor recessive; as a result, a mixture of the two gene effects is displayed.

An understanding of genetic probability is important for livestock producers to influence the quality of their herd genetics over time. Desirable traits improve animal efficiency and economic return from products. In this chapter, the Punnett square will be mastered through several genetic scenarios that need to be calculated to determine which traits will be inherited by offspring. An activity identifying common recessive and dominant traits will also demonstrate the practical use of animal genetics.

Name_____ Date_____

EXERCISE 9-1 PUNNETT SQUARE

Answer the following questions based on the information provided. Read the "Guide to Using the Punnett Square" for a reference before solving the genetic problems.

Use the following information for questions 1–3:

In dogs, the gene for fur color has two alleles. The dominant allele (F) codes for grey fur, and the recessive allele (f) codes for black fur.

1. The female dog is heterozygous. The male dog is homozygous recessive. Figure out the phenotypes and genotypes of their possible puppies by using a Punnett square.

    Genotypes
    FF: _____
    Ff: _____
    ff: _____

    Phenotypes
    Black fur: _____
    Grey fur: _____

2. The female dog has black fur. The male dog has black fur. Figure out the phenotypes and genotypes of their possible puppies by using a Punnett square.

    Genotypes
    FF: _____
    Ff: _____
    ff: _____

    Phenotypes
    Black fur: _____
    Grey fur: _____

3. The female dog is heterozygous. The male dog is heterozygous. Figure out the phenotypes and genotypes of their possible puppies by using a Punnett square.

    Genotypes
    FF: _____
    Ff: _____
    ff: _____

    Phenotypes
    Black fur: _____
    Grey fur: _____

Use the following information for questions 4–6:

In fruit flies, red eyes are dominant (E), and white eyes are recessive (e).

4. If the female fly has white eyes and the male fly has homozygous dominant red eyes, what are the possible phenotypes and genotypes of their offspring?

    Genotypes
    EE: _____
    Ee: _____
    ee: _____

    Phenotypes
    Red eyes: _____

    White eyes: _____

5. If the female fly has EE and the male fly has EE, what are the possible phenotypes and genotypes of their offspring?

   Genotypes   **Phenotypes**
   EE: _____   Red eyes: _____
   Ee: _____
   ee: _____   White eyes: _____

6. If both flies are heterozygous, then what are the possible phenotypes and genotypes of their offspring?

   Genotypes   **Phenotypes**
   EE: _____   Red eyes: _____
   Ee: _____
   ee: _____   White eyes: _____

Use the following for questions 7–9:

In dogs, there is a hereditary deafness caused by a recessive gene, "d." A kennel owner has a male dog (Gilbert) that she wants to use for breeding purposes if possible. The dog can hear.

7. What are the two possible genotypes of Gilbert?
   _____
   _____

8. If the dog's genotype is Dd, the owner does not wish to use him for breeding so that the deafness gene will not be passed on. This can be tested by breeding the dog to a deaf female (dd). Draw two Punnett squares to illustrate these two possible crosses.
   _____
   _____

9. In each case, what percentage/how many of the offspring would be expected to be hearing? Deaf? How could you tell the genotype of this male dog? Also, using Punnett square(s), show how two hearing dogs could produce deaf offspring.
   _____
   _____

A Guide to using the Punnet Square

The value of studying genetics is understanding how to predict the likelihood of an animal inheriting particular traits. This can help to develop plants and animals that have more desirable qualities.

One of the easiest ways to calculate the probability of inheriting a specific trait was invented by an early 20th century geneticist named Reginald Punnett. He developed a technique to predicting genetic traits that we now call a Punnett square. This is a simple method of discovering all of the potential combinations of genotypes that can occur in offspring, given the genotypes of their parents. It also shows us the odds of each of the offspring genotypes occurring.

Name_____ Date_____

Setting up and using a Punnett square is quite simple once you understand how it works. You begin by drawing a grid of perpendicular lines:

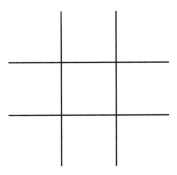

Next, you put the genotype of one parent across the top and that of the other parent down the left side. For example, if parent pea plant genotypes were YY and GG respectively, the setup would be:

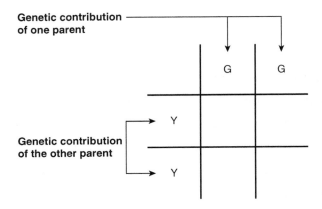

Note that only one letter goes in each box for the parents. Next, all you have to do is fill in the boxes by copying the row and column-head letters across or down into the empty squares. This gives us the predicted frequency of the potential genotypes among the offspring each time reproduction occurs.

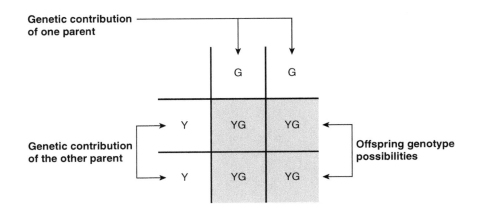

© 2016 Cengage Learning®. May not be scanned, copied or duplicated, or posted to a publicly accessible website, in whole or in part.

Name_____   Date_____

In this example, 100% of the offspring will likely be heterozygous (YG). Since the Y (yellow) allele is dominant over the G (green) allele for pea plants, 100% of the YG offspring will have a yellow Phenotype, as Mendel observed in his breeding experiments.

In the example below, if the parent plants both have heterozygous (YG) genotypes, there will be 25% YY, 50% YG, and 25% GG offspring on average. These percentages are determined based on the fact that each of the 4 offspring boxes in a Punnett square is 25% (1 out of 4). As to phenotypes, 75% will be Y and only 25% will be G. These will be the odds every time a new offspring is conceived by parents with YG genotypes.

	Y	G
Y	YY	YG
G	YG	GG

An offspring's genotype is the result of the combination of genes in the sex cells or cells that came together in its conception. One sex cell came from each parent. Sex cells normally only have one copy of the gene for each trait. Each of the two Punnett square boxes in which the parent genes for a trait are placed (across the top or on the left side) actually represents one of the two possible genotypes for a parent sex cell. Which of the two parental copies of a gene is inherited depends on which sex cell is inherited—it is a matter of chance. By placing each of the two copies in its own box has the effect of giving it a 50% chance of being inherited.

Source: Dennis O'Neil.

Name_____ Date_____

EXERCISE 9-2 DOMINANT VS. RECESSIVE ALLELE ACTIVITY

A dominant gene in a heterozygous pair hides the effect of its allele. The allele that is hidden is called a recessive gene. The polled (not having horns) condition in cattle is the result of a dominant gene and is said to be a dominant trait.

Directions: Using the word bank provided, place each trait in the appropriate column, either dominant or recessive.

Word Bank:

Black coat in horses	Nonbarred feather pattern in chickens	Albinism in animals
Color in animals	Rose comb in chickens	"Snorter" dwarfism in cattle
Red coat in cattle	Pea comb in chickens	Single comb in chickens
Normal size in cattle	Colored face in cattle	Brown coat in horses
Black coat in cattle	White face in cattle	
Barred feather pattern in chickens (also sex-linked)		

Dominant	Recessive

Name_____ Date_____

CHAPTER 9 MATCHING ACTIVITY

Term

____ 1. genetics
____ 2. genotype
____ 3. phenotype
____ 4. additive gene effect
____ 5. nonadditive gene effect
____ 6. heritability
____ 7. heritability estimate
____ 8. breeding value
____ 9. gene
____ 10. allele

Definition

a. The proportion of the total variation that is due to the additive gene effects.

b. When many different genes are involved in the expression of a trait.

c. An estimate that helps the producer to make faster genetic improvement in livestock.

d. Each gene in a gene pair, code for a different expression of the same trait.

e. expresses the likelihood of a trait being passed from one parent to offspring.

f. The study of heredity.

g. these traits are readily observable and are controlled by only one or a few gene pairs.

h. A unit of heredity that is transferred from a parent to offspring.

i. The combination of genes that an individual possesses.

j. The physical appearance of an animal.

Name_____ Date_____

CHAPTER 9 LAB QUESTIONS

1. Who was Gregor Mendel? What was his contribution to the study of genetics?

2. Who was Reginald Punnett? What was his contribution to the study of genetics?

3. How does the Punnett square work? Explain.

4. What are hereditability estimates used for?

Chapter 10
Animal Reproduction

INTRODUCTION

Through mating, female and males animals begin the intricate process of reproduction. The male gamete, called the sperm, is deposited in the female reproductive tract, where it moves to the oviduct and fertilizes the egg cell (female gamete). In livestock, all the females go through a period during which they will accept the male for mating; this is called the estrus cycle. During estrus, otherwise known as the heat period, hormones cause the egg cell to be released for fertilization.

If the egg cell is fertilized, it moves to the uterus, where it grows into a fetus; pregnancy is maintained by hormones and body functions. When the fetus has developed for a full term and is ready for birth, the uterine muscles contract, forcing the fetus through the birth canal. Reproduction in poultry is similar in some ways to reproduction in mammals. The main difference is that the embryo develops in the eggshell outside of the mother's body. Eggs must be kept at the proper temperature and humidity to hatch. This is usually done in an incubator.

On many livestock farms, animals are raised in separate groups depending on their reproductive development. If animals are placed in the wrong reproductive group, many negative outcomes are possible: (1) animals may be stressed and lower in daily weight gain or production, (2) animals that should not be bred due to health problems are at risk, (3) young animals can die from premature pregnancies if placed in the wrong group, and (4) young males can be stressed and injured from aggressive older livestock in the same group.

When managing livestock reproduction, knowledge of the anatomy is very important, as well as understanding the normal functions of the reproductive system during pregnancy and birth. If an animal has difficult labor, a manager must be able to recognize the problem and assist with the birth without harming with the cow or the calf. In this chapter, you will practice grouping animals at different reproductive stages of life, as well as identifying reproductive anatomy parts.

Name_____ Date_____

EXERCISE 10-1 REPRODUCTIVE PLANNING FOR LIVESTOCK

A beef cattle operation must manage the reproductive cycles and interactions between male and female animals on the farm. When raising cattle, knowledge of the reproductive development is a key factor to select animals for breeding purposes. A producer must be able to identify the stage of reproductive development that animals are at and separate livestock based on the breeding strategy. If a cow is "open," it means it is not pregnant

Following is a mixed group of cattle at different stages of development. Each animal will need to be separated into the appropriate group for development. Study the herd records, and identify which animals belong in each group of reproductive development. Place the proper animal identification number in the group it should be integrated into.

Mixed Cattle Group: Herd List

Identification Number	Gender/Age	Animal Status
1012	Male/24 months	Castrated/weighs 1,200 lb
9000	Female/52 months	Open/2-month-old calf at side
2611	Female/7 months	Open/650 lb
1029	Male/10 months	Uncastrated/900 lb
8826	Female/46 months	Open/1-week-old calf at side
2020	Female/120 months	Open last two breeding cycles
1893	Female/12 months	Open/weighs 950 lb
1982	Male/18 Months	Castrated/weighs 1,100 lb
1729	Female/24 months	8 months pregnant
2001	Female/36 months	5 months pregnant
4233	Female/9 months	Open/weighs 800 lb
1482	Female/36 months	Open/aborted last year's calf
1673	Male/48 months	Uncastrated herd sire

Name_____ Date_____

8675	Male/7 months	Castrated/weighs 700 lb
9212	Female/14 months	Open/weighs 950 lb
3420	Male/7 months	Uncastrated/750 lb
4200	Female/95 months	3 months pregnant

Directions: Place the individual cattle into the appropriate groups. List the identification number and provide a reason that animal was placed in this group.

Bulls that are old enough for breeding and steers that are over 800 pounds:
_____
_____

Cattle that are too old for reproduction:
_____
_____

Cattle that are too young or small for breeding:
_____
_____

Cattle that need to be bred and are the correct age and weight:
_____
_____

Cattle that are of age and size, but should not be bred currently:
_____
_____

Name_____ Date_____

EXERCISE 10-2 REPRODUCTIVE ANATOMY ACTIVITY

Special Instructions: Reproductive tracts can be obtained from a local slaughterhouse or possibly a livestock producer. Class members performing the dissection should use caution because the scalpel can become slippery during the dissection. All participants should wear gloves as protection against any disease organisms that may be present. Using the following anatomy diagrams (Figures 10-1 through 10-5), identify the parts of each sample reproductive tract.

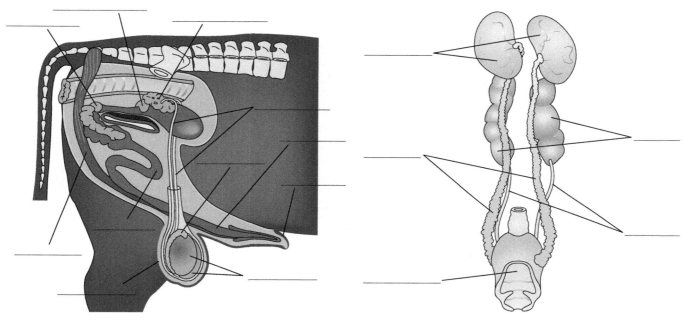

Figure 10-1 Reproductive organs of a bull.

Figure 10-2 Reproductive system of the male chicken.

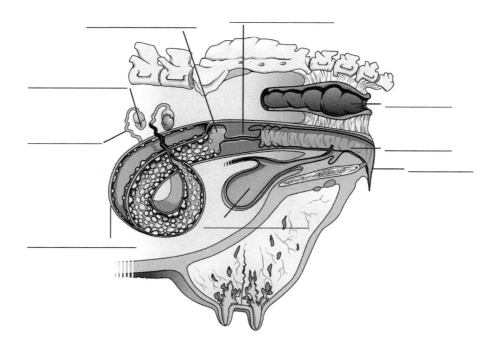

Figure 10-3 Female reproductive system of a cow, side view.

94 SECTION 3 Animal Breeding

Name_____ Date_____

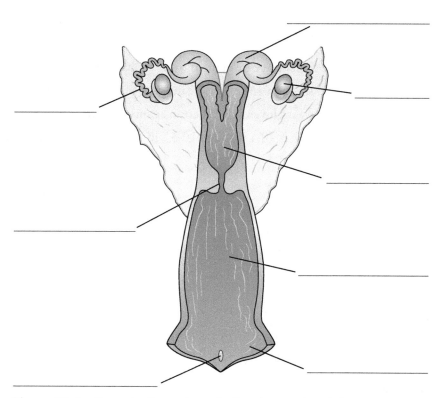

Figure 10-4 Reproductive system of a female cow, dorsal view.

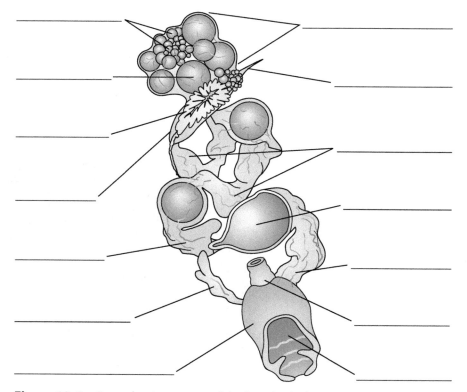

Figure 10-5 Reproductive system of the female chicken.

© 2016 Cengage Learning®. May not be scanned, copied or duplicated, or posted to a publicly accessible website, in whole or in part.

Name_____ Date_____

Part 1: Female Anatomy

Materials needed:

    Reproductive tract from cow, ewe, and sow

    Surgical gloves for each class member

    Scalpels

    Dissecting pins and labels

    Plastic aprons for each class member

Procedure:

1. Lay tracts from different species side by side so that comparisons can be made.
2. Prior to dissecting the tracts, identify the ovaries, infundibulum, oviducts, uterine horns, uterine body, cervix, vagina, and vulva. Note differences between species. Label parts with dissecting pins.
3. Identify any structures on the ovaries.
4. Dissect the uterus of the cow and/or ewe and note the septum that divides what, from the outside, appears to be the uterine body.
5. Dissect the cervix of each species. Identify rings or folds in the cervix. Note differences between species.
6. Dissect the vagina of each species. Locate the urethral opening in the floor of the vagina. Locate the external os at the junction of the cervix and vagina. Compare the external os of the cow and/or ewe to the external os of the sow. Identify the clitoris, which is located at the floor of the vagina where the vagina joins the vulva. Label parts with dissecting pins.
7. Locate semen deposition sites in each species for natural service and artificial insemination. Label with dissecting pins.

Part 2: Male Anatomy

Materials needed:

    Reproductive tracts from bull, ram, and boar

    Surgical gloves for each class member

    Scalpels

    Dissecting pins and labels

Procedure:

1. Lay tracts from different species side by side so that comparisons can be made.
2. Identify the scrotum, testicle, epididymis, vas deferens, urethra, Cowper's gland, prostate, seminal vesicles, and penis. Locate the sigmoid flexure of the penis. Label all parts with dissecting pins.
3. Dissect one testicle and locate the ducts of the rete testes. Label with dissecting pins.

Name_____ Date_____

CHAPTER 10 MATCHING ACTIVITY

Term   **Definition**

____ 1. reproduction

____ 2. copulation

____ 3. embryo

____ 4. parturition

____ 5. scrotum

____ 6. sterile

____ 7. ridgeling

____ 8. vas deferens

____ 9. gestation

____ 10. estrogen

a. the act of giving birth; the final step in reproduction

b. cannot produce live sperm

c. period of the time during which the animal is pregnant

d. sexual reproduction beginning with the mating of the male and female

e. a male in which one or both testicles are held in the body cavity

f. a tube that connects the epididymis with the urethra.

g. saclike part of the male reproductive system outside the body cavity that contains the testicles and the epididymis

h. when organisms multiply or produce offspring

i. female sex hormones

j. fertilization that occurs when the sperm penetrates the egg cell and a new animal begins to grow

Name_____ Date_____

CHAPTER 10 LAB QUESTIONS

1. Explain estrus and how it functions.

   _____
   _____
   _____
   _____
   _____
   _____

2. Describe the problems that can arise during livestock parturition.

   _____
   _____
   _____
   _____
   _____
   _____

3. Why is it important that animals be grouped according to reproductive status?

   _____
   _____
   _____
   _____
   _____
   _____

Chapter 11
Biotechnology in Livestock Production

INTRODUCTION

Since early human civilization, humans have been using some forms of biotechnology to change living plants and animals for commercial use. Activities as simple as livestock breeding, crop improvement, and the production of food products are basic forms of biotechnology. As microbiology and technology increased in the 1970s, laboratory techniques were developed that allowed researchers to identify and manipulate the DNA found in cells of living organisms. The DNA contains the information about an organism's genetic makeup. As this technology has improved over the past 40 years, the ability to influence many characteristics of living organisms has increased. The science of altering genetic and reproductive processes in animals and plants is called agricultural biotechnology.

Most of the work being done in research and biotechnology is currently focused on genetic engineering and embryo transfer. Genetic engineering involves taking a tiny bit of DNA containing the desired gene from one organism and splicing it onto the DNA strand in another organism. The recipient organism takes on the characteristic controlled by the transferred gene. There are many forms of genetically engineered techniques and substances that can be administered to animals to affect their performance or characteristics. Many other areas of research are currently being conducted with genetic engineering in animal and plant

Protestors against the use of genetically modified products in agriculture.

science. One example of genetic engineering is Paylean®, ractopamine hydrochloride, which increases feed efficiency, encourages growth in muscles, and lowers the fat content in hogs.

Another example of genetic engineering in animals is bovine somatotropin (bST) in dairy cows, which results in higher milk production. Bovine somatotropin causes energy derived from feed to be used for milk production rather than for weight gain, but it does not reduce the energy available for body maintenance. Small amounts of bST are produced naturally, but the amount of bST produced by dairy cows has gradually increased over the years as a result of breeding selection. Through the use of genetic engineering, large quantities of bST can be produced in an injectable form. The manufactured hormone is called recombinant bovine somatotropin (rbST). The gene that controls bST production is spliced into the DNA of bacteria, which is then injected into the cow. Consumers have expressed concern over the increase in hormone levels with the use of rbST and the chance that elevated levels of hormones could be transferred to humans that consume milk from a cow that was injected with rbST. Currently, some milk products are labeled as "not containing rbST" in order to give consumers an alternative.

Another major tool from biotechnology is embryo transfer. This has become an established technology in cattle production. In embryo transfer, the embryos from an animal with desirable genetic characteristics are "flushed" out of the animal; the embryo is fertilized and then implanted in a surrogate animal to carry out the pregnancy. The use of embryo transfer permits the production of many more offspring from genetically desirable animals.

Genetically modified crops (GMOs) are very beneficial in terms of increased production of agricultural products in a world with a growing population, but the long-term impacts of this recent technology cause concern in consumers and activists. In this chapter, the impacts of using of rbST are explored, and a debate on the use of genetically modified crops is conducted.

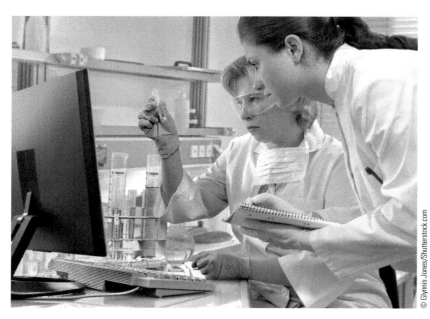

Scientists researching new ways to utilize biotechnology.

Name_____ Date_____

EXERCISE 11-1 IMPACTS OF USING rbST

Directions: Using the information in the textbook and this chapter, make a list of positive and negative effects of using rbST in dairy cattle. Once finished, come to a conclusion that includes your own opinion about the use of rbST.

Positive Effects of Using rbST	Negative Effects of Using rbST

Conclusion: How does this information affect your buying decision as a consumer of milk? Does this affect animal welfare at all? Give your "educated" opinion in detail.

Name_____ Date_____

EXERCISE 11-2 GENETICALLY MODIFIED ORGANISMS DEBATE ACTIVITY

Pre-debate Exercise

Directions: First research and read two articles about GMOs using the Internet or library resources. One article should be supportive of using GMO technology in our food system and the other should be against the use of GMOs. After reading, complete the following questions and tasks using facts and arguments found in the readings and/or from your prior knowledge.

1. What are genetically modified crops?
   _____
   _____
   _____

2. Make a list of positive effects of genetically modified crops.
   _____
   _____
   _____

3. Make a list of negative effects or risks associated with genetically modified crops.
   _____
   _____
   _____

4. How do genetically modified crops affect the hunger crisis in the world?
   _____
   _____
   _____

5. In New York, should products in the store be labeled if they have genetically modified ingredients? Explain why or why not.
   _____
   _____
   _____

6. Is there a way to supply food to the world without using GMO products?
   _____
   _____

Name_____ Date_____

Four-Corner Debate Activity

When defending an opinion during a debate, one must study both sides of an argument in order to demonstrate a true knowledge of the topic and establish credibility for an opinion.

Directions: In the classroom, each of the four corners will have a sign posted designating an opinion. The signs will read (1) agree, (2) strongly agree, (3) disagree, and (4) strongly disagree. When the instructor reads a controversial statement about GMOs, students must decide which sign best represents their opinion and go stand next to the sign. The instructor will then ask each student to defend his or her opinion with a well-educated explanation.

Name_____ Date_____

CHAPTER 11 MATCHING ACTIVITY

Term	Definition
____ 1. biotechnology	a. proteins that affect the utilization of energy in the body
____ 2. transgenic	b. the science of altering genetic and reproductive processes in animals and plants
____ 3. clone	
____ 4. genetic engineering	c. by utilizing this technology, it is possible to produce many more offspring from desirable animals than would be possible if the animals had to carry each embryo to term before producing another one
____ 5. somatotropin	
____ 6. bovine somatotropin	
____ 7. recombinant bovine somatotropin (rbST)	d. the process in which the gene that controls bST production is spliced into the DNA of bacteria, which is then injected into the cow
____ 8. superovulation	e. when cells or organisms are genetically identical to each other
____ 9. agricultural biotechnology	f. manipulating the DNA of living organisms at the molecular and cellular level to produce new commercial applications
	g. causes energy derived from feed to be used for milk production rather than for weight gain, but does not reduce the energy available for body maintenance
	h. the process in which human genes are inserted into animal embryos
	i. the process of identifying and transferring a gene or genes for a specific trait from one organism to another

Name_____ Date_____

CHAPTER 11 LAB QUESTIONS

1. Describe some of the ways biotechnology is now being used in agriculture.

   _____
   _____
   _____
   _____
   _____
   _____

2. How does genetic engineering work? Describe the process.

   _____
   _____
   _____
   _____
   _____
   _____
   _____

3. How is rbST used as a livestock management tool?

   _____
   _____
   _____
   _____
   _____
   _____

4. How is embryo transfer used as a livestock management tool?

   _____
   _____
   _____
   _____
   _____
   _____

Chapter 12
Animal Breeding Systems

INTRODUCTION

Strategic livestock breeding can be beneficial in many ways by improving herd quality and performance factors, as well as by making the herd more economically efficient. The two basic systems of livestock breeding are straightbreeding and crossbreeding. The type of breeding system used depends largely on the farmer and the goals of the operation. Some livestock operations choose to maintain purebred animals for various reasons. Purebred animals can be eligible for registry in a breed association with documentation of the bloodline. Establishing a purebred herd takes intense management and investment in pure pedigree animals.

Purebred animals are not always the most advantageous for agricultural production. Sometimes, a crossbred animal will contain a combination of more desirable traits than a purebred could achieve. Many livestock producers choose to manage a crossbreeding system where two different breeds are combined to achieve the desired traits; the resulting offspring are called a hybrid. Superior traits that result from crossbreeding are called hybrid vigor, which is measured by the average superiority of the hybrid offspring over the average of the parents.

Research has shown that well-planned crossbreeding programs can increase total productivity in beef herds by 20 to 25 percent. In order to achieve this type of increase in productivity, the producer must follow a good performance selection program, good management, good nutrition, and good herd health practices to achieve the desired results. A majority of livestock producers use crossbreeding, except for most dairy producers and many breeds of horses.

Crossbreeding systems range from those that are very complex to those that are very simple, so each producer needs to assess which system works best for the current circumstances of the operation. Some of the common systems used are terminal sire crossed with F_1 females, rotating the herd bull every 3 or 4 years, two-breed rotation, three-breed rotation, four- and five-breed rotations, static terminal sire systems, rotational-terminal sire system, and composite breeding systems. In this chapter, Exercise 12-1 provides examples of livestock operations that have a desired trait that they are trying to attain through breeding strategies. For each example, use the breeding techniques discussed in the textbook to suggest the proper breeding strategy to achieve the desired hybrid vigor.

Linebreeding
(A represents the male; B and C represent females)

1st mating: A X B   A X C
1st generation: ½ A and ½ B   ½ A and ½ C

2nd mating: ½ A and ½ B X ½ A and ½ C
2nd generation: ½ A and ¼ B and ¼ C

The offspring in the second generation have received 50 percent of their genetic inheritance from the sire A because he appears twice in their pedigree. They have received only 25 percent of their genetic inheritance form each of the females B and C.

Grading Up
(A_1, A_2, A_3 represent purebred sires of a given breed; G represents a grade female)

1st mating: A_1 X G
1st generation: ½ A_1 and ½ G (50% purebred, 50% grade)

2nd mating: A_2 X ½ A_1 and ½ G
2nd generation: ½ A_2 and ¼ A_1 and ¼ G (75% purebred, 25% grade)

3rd mating: A_3 X ½ A_2 and ¼ A_1 and ¼ G
3rd generation: ½ A_3 and ¼ A_2 and ⅛ A_1 and ⅛ G (87.5% purebred, 12.5% grade)

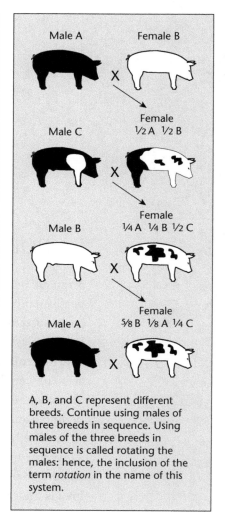

A, B, and C represent different breeds. Continue using males of three breeds in sequence. Using males of the three breeds in sequence is called rotating the males: hence, the inclusion of the term *rotation* in the name of this system.

Name_____ Date_____

EXERCISE 12-1 IMPROVING HERD QUALITY THROUGH PROPER BREEDING STRATEGY

In most livestock used for meat products, crossbreeding can result in a combination of many desirable traits and long-term herd improvement. Controlled crossbreeding takes a high level of management and record keeping. This exercise gives example swine herd scenarios and the desired combinations of breeds and traits that the farmer is trying to produce through crossbreeding strategies.

Directions: For each example herd scenario, show the combinations needed to use the existing herd and make the offspring a combination of desired breeds. All offspring should have the desired percentage of each breed. Make sure to show all work when explaining crossbreeding strategies.

Scenario 1

Herd 1: A dairy farmer has a purebred herd of Holsteins. Some of the cows produce a high level of milk every day (A), and another group of cows have good conformation and good feet and legs (B). The farmer has also just bought a high-pedigree bull (C) that has proven to have a good calving ease trait in his offspring. The dairy farmer hopes to use a linebreeding strategy to combine these traits over four generations of offspring.

Suggest the needed breeding combinations, and determine the percentage of inherited genetics in the fourth generation that must come from each of the first animals to be mated. Show all work:

Scenario 2

Herd 2: A beef cattle rancher has a herd of grade females that are not purebred. The farmer decides to use breeding strategy to improve his herd. The national price for beef from cows that are 90 percent angus breed are more valuable than grade animals. The farmer plans to buy a new purebred angus bull to sire one generation of offspring. A new bull will be introduced to the bull every generation for four generations. The letters A, B, C, and D represent the purebred angus sires that will be used to help the beef operation grade up. G represents a grade female.

Name_____ Date_____

Demonstrate the change that will occur in the beef through four generations of using the grade-up strategy. Show in fractions the amount of hereditary traits in the offspring that result from each of the original mated animals. Show all work:

Scenario 1

Herd 3: A hog farmer wants to improve the growth and performance of the herd and decides to use a three-breed rotation crossbreeding system. The farmer decides to use two sows from the herd to test this new breeding system. One of the females is a white Heritage breed, and the other female is a Yorkshire breed. The farmer will breed each of the sows with three different boars over three generations. The first generation will be sired by a Tamsworth boar, the second generation will be sired by a Duroc boar, and the third generation will be sired by a Berkshire boar.

Demonstrate the combination of genetics that will take place in each generation for both the Heritage sow and the Yorkshire sow. Show all work:

Name_____ Date_____

CHAPTER 12 MATCHING

Term

____ 1. straightbreeding
____ 2. purebred
____ 3. inbreeding
____ 4. closebreeding
____ 5. linebreeding
____ 6. outcrossing
____ 7. linecrossing
____ 8. grading up
____ 9. grade
____ 10. hybrid

Definition

a. the animals being mated are very closely related and can be traced back to more than one common ancestor

b. the mating of purebred sires to grade females

c. mating animals from two different lines of breeding within a breed

d. the resulting offspring from crossbreeding

e. an animal of a particular breed that has the characteristics of the breed to which it belongs

f. any animal not eligible for purebred registry

g. the mating of related animals

h. the mating of animals that are more distantly related

i. the mating of animals of the same breed

j. the mating of animals of different families within the same breed

Name_____ Date_____

CHAPTER 12 LAB QUESTIONS

1. What is the difference between straightbreeding and linebreeding?

   _____
   _____
   _____
   _____
   _____
   _____

2. Describe each of the commonly used livestock breeding systems.

   _____
   _____
   _____
   _____
   _____
   _____
   _____

3. What are the benefits that result from crossbreeding livestock?

   _____
   _____
   _____
   _____
   _____
   _____

4. What are the factors that affect which breeding system a farmer chooses to implement?

   _____
   _____
   _____
   _____
   _____
   _____

SECTION 4
Beef Cattle

Chapter 13	Breeds of Beef Cattle	112
Chapter 14	Selection and Judging of Beef Cattle	117
Chapter 15	Feeding and Management of the Cow-Calf Herd	124
Chapter 16	Feeding and Management of Feeder Cattle	132
Chapter 17	Diseases and Parasites of Beef Cattle	138
Chapter 18	Beef Cattle Housing and Equipment	141
Chapter 19	Marketing Beef Cattle	150

Chapter 13
Breeds of Beef Cattle

INTRODUCTION

Many of the beef breeds used in modern agriculture had their origins in Europe. In the latter part of the 1700s, selection of cattle to form breeds began mainly in the British Isles. The cattle were selected for the most desirable traits for the area. The most desirable animals were kept as breeding stock. The farmer then removed those animals that did not have the desired traits from the breeding herd. The practice of removing animals from the herd is referred to as culling. This method was repeated to increase gene frequency, and the early European farmers rarely used the technique except with breeding stock. As these purebred animals became popular, many breed registries were formed to establish the origins of the breed. Since then, all animals registered to the breed must trace the ancestry back to the original animals in the breed registry.

In the United States, some newer hybrid breeds have been developed as a result of selecting crossbred animals that possessed desirable traits. An example of this practice is the Brahman breed. Some other breeds which have been introduced in the United States are Simmental, Limousin, Blonde d'Acquitaine, Chianina, Maine-Anjou, Charolais, and several others. There are more than 50 breeds of beef cattle available to producers in the United States. Each breed has specific characteristics and origins that make it desirable for producers. Some points that should be considered when selecting a breed include the following: all breeds should have both strong and weak traits, there is no single breed is best for all traits, and every breed has a wide range of genetic variation. A farmer must keep in mind that when selecting animals as breeding stock, the use of good breeding practices has more impact than the particular breed selected. The producer needs to select a breed that seems to produce well in the area where it will be raised, and it should have proven market demand.

In this chapter, the focus will be on selecting beef cattle based on the specific goals of the cattle operation. Demonstrate a knowledge of the beef breeds and practice breed selection in Exercise 13-1.

Name_____ Date_____

EXERCISE 13-1 BREED SELECTION ACTIVITY

When establishing a beef operation, a farmer must create goals for the farm and then decide which breed or crossbreed is best suited for the operation. For this activity, assume that the goals of the operation are (1) to have a low cost per unit of weight gain, (2) to have polled livestock to reduce herd injuries, (3) to produce a high-quality, well-marbled carcass, and (4) to have cows that will have a good mothering ability.

Directions: **Part 1:** Each of the following pictures show a different beef breed. For each picture, identify the breed and write a brief history and description of the breed. Include any unique characteristics of the breed. **Part 2:** Decide which breed or cross of breeds will meet the goals of the beef operation. Explain why the selected breed was the most advantageous.

Part 1

1.

   a. Breed: _____
   b. History: _____
   c. Description: _____

2.

   a. Breed: _____
   b. History: _____
   c. Description: _____

3.

   a. Breed: _____
   b. History: _____
   c. Description: _____

4.

   a. Breed: _____
   b. History: _____
   c. Description: _____

Name_____ Date_____

5.

   a. Breed: _____
   b. History: _____
   c. Description: _____

6.

   a. Breed: _____
   b. History: _____
   c. Description: _____

Part 2

Make a decision of which breeds or combination of breeds will meet the goals of the beef operation. Explain the decision in detail.

Name_____ Date_____

CHAPTER 13 MATCHING ACTIVITY

Term

____ 1. vealers
____ 2. dewlap
____ 3. Angus
____ 4. Barzona
____ 5. Beefmaster
____ 6. Blonde d'Aquitaine
____ 7. Braford breed
____ 8. Brahman
____ 9. Charolais
____ 10. Devon cattle

Definition

a. originated in 1931 in Texas
b. one of the oldest of the French breeds of beef cattle
c. originated in 1961 when several French breeds were combined
d. created when strains of the Zebu cattle were bred to females from several British breeds of cattle
e. In 1862, the first herd book of this breed was published.
f. Some authorities believe they descended from *Bos longifrons,* a small type of aboriginal cattle in Britain.
g. calves grown for veal
h. developed in Arizona beginning in 1942
i. loose skin under the throat
j. began in 1947 by crossing Hereford bulls on Brahman cows

Name_____ Date_____

CHAPTER 13 LAB QUESTIONS

1. When and where did beef cattle breed selection begin?

2. What considerations are made when selecting breeds of cattle for a beef operation?

3. How many beef breeds are there currently in the United States?

4. Name three more recent breeds that have been developed in the United States.

Chapter 14

Selection and Judging of Beef Cattle

INTRODUCTION

Once a producer has identified a breed to raise, the next step is selecting cattle with quality traits that will be passed on to each generation of the herd. The three main kinds of beef cattle production systems are (1) cow-calf producers, (2) purebred breeders, and (3) cattle feeders. Each of these systems requires specific cattle characteristics to improve functionality of the operation. Cow-calf producers keep herds of cattle and produce calves for sale to slaughter cattle producers, so calving ease and calf growth are important characteristics. Purebred producers furnish breeding stock to cow-calf producers; breeding stock should contain a combination of many genetic qualities. Cattle feeders buy calves or yearlings and feed them to market weights. They base their selection on body condition, muscling, and rate of gain.

Performance records help the producer select animals based on records of previous performance. These traits that are measured for performance are thought of as hereditable; therefore, a good performance score will increase the genetic value of the animal. Some traits are more heritable than others, so a cattle producer must use the knowledge to improve the beef herd.

Beef animals are judged on their conformation and are given a score based on the visible body characteristics. Many county fairs and cattle shows have judging events, followed by a cattle sale. Producers present their animals in a show ring to be judged in a class of similar cattle from other producers. The more judging events a beef animal wins or performs well in, the more valuable its genetics will be. A good judge learns the parts of the animal and develops a logical system for evaluating the animal. A judge should have experience and credibility in the cattle industry in order to build a reputation among cattle producers. Once a judge places a class of cattle in order from first to last, the judge should be prepared to explain why the cattle were selected in that order. The judge should give a presentation of oral reasons and use the proper terms to describe cattle conformation. There are cattle judging competitions in which young judges test their ability to properly select cattle based on conformation.

This chapter provides the opportunity to practice body condition scoring that is useful for feeder cattle producers, as well as an exercise to prepare oral reasons for a class of beef cows.

Name_____ Date_____

EXERCISE 14-1 BODY CONDITION SCORING ACTIVITY

A body condition score (BCS) describes the degree of fatness of an animal. Animals of similar weights may differ dramatically in BCS, so a score is given based on fat cover in the brisket; on the ribs, back, hooks, and pins; and around the tailhead. A body condition scoring system has been established and provides a numerical range of 1 to 9 to identify varying degrees of fat cover on beef cattle.

BCS	Description
1	Severely emaciated. Little evidence of fat deposits or muscling.
2	Emaciated. Little evidence of fat but some muscle in hindquarters.
3	Very thin. No fat on ribs or brisket. Backbone easily visible.
4	Thin, with ribs easily visible but shoulders and hindquarters still showing fair muscling. Backbone visible.
5	Moderate to thin. Last two or three ribs cannot be seen unless animal has been shrunk. Little evidence of fat in brisket, over ribs, or around tailhead.
6	Good, smooth appearance. Some fat deposits in brisket and over tailhead. Ribs covered, and back appears rounded.
7	Very good flesh, brisket full. Ribs very smooth.
8	Obese, back very square, heavy fat pockets around tailhead. Square appearance.
9	Rarely observed. Very obese. Mobility may be impaired by excessive fat.

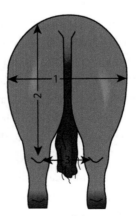

Things to Look For:
1. Width of round
2. Depth of round
3. Width between legs

Name_____ Date_____

Directions: For each of the following beef cows, determine the body condition score of the animal and the reasons it was placed in this range.

MUSCLE THICKNESS

No. 1
No. 2
No. 3
No. 4

No. 1, No. 2, and No. 3 thickness pictures depict minimum grade requirements. The No. 4 picture represents an animal typical of the grade.

1. BCS:
_____
_____

Explain:
_____
_____
_____

2. BCS:
_____
_____

Explain:
_____
_____
_____

Name_____ Date_____

3. BCS:
_____
_____

Explain:
_____
_____
_____

4. BCS:
_____
_____

Explain:
_____
_____
_____

Name_____ Date_____

EXERCISE 14-2   GIVING ORAL REASONS

A cattle judging is practiced in 4-H events, FFA competitions, and even collegiate-level cattle judging competitions. For this exercise, the instructor will provide access to a group of show cattle to be judged. Use a group of approximately six animals for this judging exercise. While judging, make sure to walk around the cattle to see all the areas of interest for conformation and have cattle move around the show ring as much as necessary for evaluation. Oral reasons are given to explain the differences between animals that influenced their placing.

Use the following procedure to properly prepare to evaluate and give oral reasons for the group of cattle being evaluated.

1. Use proper terms that describe the characteristics of the animals. Beef cattle judging terms are listed in the textbook; use these terms when describing the differences in cattle.

2. When giving your reasons, stand 6 to 8 feet from the judge. Maintain eye contact and use good body posture during the presentation of reasons. Reasons are given in a logical order and should be well organized. Speak clearly and loudly enough to be heard. Speak with confidence, and make a concise final statement explaining why the animal at the bottom of the class was so placed. Do not speak for more than 2 minutes total during reasons.

3. Keep accurate notes

4. The evaluator should write notes in order to give accurate reasons. A set of reasons is organized in the following way:

   Introduction—Give the name of the class and how it was placed.

   Top pair—Give the reasons for placing 1 over 2, using comparative terms. Grant the advantages of 2 over 1, also using comparative terms. Finally, criticize 2 using either descriptive or comparative terms.

   Middle pair—Follow the same outline used for the top pair.

   Bottom pair—Follow the same outline used for the top pair.

   Summary—Give a final statement that explains why the animal at the bottom of the class was so placed. Mention the animal's strong points, if any, and describe its major faults.

Class #3: Market Steers			Placing: 3-1-2-4	
1 Black		2 Hereford	3 Black Baldy	4 Red
3/1	Heavy muscled More fat cover Apt to grade choice More lbs. of product Heavy boned		1/3 Grant Higher cutability Moderate frame	
			1 (Fault) Poor balance Over-finished	
1/2	Better off hind two legs Heavy muscled Market-ready look Wider chested		2/1 (Grant) Shapely, more expressive Clean lines Level hip	
			2 (Fault) Short coupled Narrow made	
2/4	Harvest with a more shapely carcass More even fat deposition		4/2 (Grant) Extended Structurally sound Deep bodied	
			4 (Fault) Light muscled Under finished Requires more days to grade choice Fine boned	

CHAPTER 14 MATCHING ACTIVITY

Term

____ 1. cow-calf system
____ 2. feeder calf
____ 3. frame score
____ 4. conformation
____ 5. production testing
____ 6. progeny testing
____ 7. performance testing
____ 8. pedigree
____ 9. sire summary
____ 10. expected progeny difference (EPD)

Definition

a. refers to the appearance of the live animal; includes the skeletal structure, muscling, fat balance, straightness of the animal's lines, and structural soundness

b. the record of the ancestors of an animal

c. defined as a method of collecting records on beef cattle herds to be used for selecting the most productive animals

d. weaned calves that are under 1 year of age and are sold to be fed for more growth

e. provides information on traits that are economically important to cattle producers; included is information regarding the ability of the bull to transmit growth rate to his offspring

f. a herd of cows that are bred each year to produce calves.; the calves are then sold to cattle feeders, who feed them to slaughter weights

g. a measure of the degree of difference between the progeny of the bull and the progeny of the average bull of the breed in the trait being measured

h. refers to measuring a brood cow's production by the performance of its offspring

i. usually refers to the evaluation of a bull by the performance of a number of its offspring

j. a measurement based on observation and height measurements when calves are evaluated at 205 days of age

Name_____ Date_____

CHAPTER 14 LAB QUESTIONS

1. What are the three main types of beef cattle operations? Describe how each operation works.

2. How is a beef cow's condition measured?

3. Why are cattle judged? What measurements are the placings based upon?

Chapter 15

Feeding and Management of the Cow-Calf Herd

INTRODUCTION

In order to properly manage a beef cow operation, a producer must analyze all aspects of the business. Integrated resource management (IRM) is a management tool available to beef producers that uses a team approach to improve the competitiveness, efficiency, and profitability of their beef business; this system is mainly utilized by large operations. A typical team consists of accountants, bankers, extension personnel, veterinarians, university researchers, soil and water conservationists, and people from allied industries. The team analyzes the beef production practices of the farm or ranch to diagnose and solve inefficiencies and make sure the production activities complement each other. Both production and financial performance data are gathered and analyzed using the standard performance analysis (SPA). These data are used in all aspects of the business to control the amount of investment in cattle in relation to the returns. When feeding cattle in a beef operation, accurate calculation of feeds given at various stages of development is important.

Feeding programs for cow-calf beef herds are based on the use of roughages. Roughages are commonly provided by pasture in the summer and silage and hay in the winter. The kinds of pasture, silages, and hay used depend on the climate, soil, and land make-up of the geographic area. When preparing hay materials for cattle, dry hay can be produced in small or large square bales or, more commonly, in the form of round bales. Each category of beef animal has different feed requirements based on their stage of development. Dry, pregnant cows are fed to prevent their becoming too fat or too thin during the winter. Younger cows and heifers require more feed than mature cows. Breeding bulls also have periods of high-energy requirements, such as during the breeding season. Calves may be creep fed commercial pelleted grain when being marketed after weaning.

Bull calves to be kept for breeding purposes are weaned at 6 to 8 months of age. High-energy rations are fed for 5 months to determine which bulls have the best gaining ability. Bulls that gain best should be selected for breeding, and the others should be marketed for slaughter.

The bulls that are kept should begin breeding at 15 to 18 months of age. With the proper breeding strategy, calving should occur in a 40- to 60-day period. Keeping births within this time period gives calves a more uniform weight and age. Calving should be timed out to occur 6 weeks to 3 months before pasture season, depending on local climate.

During calving, farmers should monitor the cows closely for signs of parturition. Most cattle on pasture will not need assistance in delivery, but many young heifers may need help during calving. After birth, calves should be maintained with common herd health practices. Castration, dehorning, vaccination, and identification of calves are best done when they are young so there is less shock to the calf since it has not yet developed all of its sensory neurons. Some farmers who do not own cow herds may buy calves at weaning and prepare them for the feedlot, where they will be finished and sent to the slaughterhouse.

This chapter analyses the feeding requirements of beef cattle at different stages of production in Exercise 15-1. Exercise 15-2 helps to develop proper calf care practices.

Feed bunk on a large feedlot.

Name_____ Date_____

EXERCISE 15-1 FEEDING COWS AND CALVES AT DIFFERENT STAGES OF GROWTH

Maintaining feed with the proper available energy for the requirements of beef cattle at different stages of growth is imperative efficient management. Producers try to avoid cattle losing too much condition or gaining too much condition based on the stage of development of each animal.

Directions: For each of the following categories of beef cattle, describe the suggested ration for proper body condition maintenance. (*Hint:* Suggested rations are provided in the textbook.)

1. **Creep Feeding Calves—Self-Fed (100-lb mix)**

   _____
   _____
   _____
   _____
   _____

2. **Feed Mixtures for Weaned Bull Calves (weaning to about 700 lb)**

   _____
   _____
   _____
   _____
   _____

3. **Feed Mixtures for Bulls over 700 lb**

   _____
   _____
   _____
   _____
   _____

4. **Wintering Rations for Bred Heifers (800 to 900 lb)**

   _____
   _____
   _____
   _____
   _____

Name_____ Date_____

5. **Rations to Grow Replacement Heifers (450–500 lb; weight gain 1–1.25 lb daily)**

_____
_____
_____
_____
_____

6. **First-Calf Heifers—Lactation Rations—Drylot**

_____
_____
_____
_____
_____

7. **Lactating Rations for Cows in Drylot (1,100 lb)**

_____
_____
_____
_____
_____

8. **Wintering Dry, Pregnant Beef Cows (1,000 to 1,100 lb)**

_____
_____
_____
_____
_____

128 SECTION 4 Beef Cattle

Name_____ Date_____

EXERCISE 15-2 CALF CARE ACTIVITY

Calf care begins from the time that a cow begins the calving process. A cattle manager must understand the signs and steps in the calving process to assist in a safe and healthy calving experience. Once the calf is born, several immediate practices are necessary to help the calf to survive the first 48 hours of life. Once a calf has passed that threshold, other management practices are needed to prepare the calf for the purposed of the beef operation.

Directions: For each of the calf care practices mentioned, describe the proper implementation of the practice in detail.

Calving practices and procedures:

_____
_____
_____
_____
_____

First 48 hours after calving:

_____
_____
_____
_____
_____

Castration:

_____
_____
_____
_____
_____

Dehorning:

_____
_____
_____
_____
_____

Name_____ Date_____

Branding and marking:
_____
_____
_____
_____
_____

Weaning:
_____
_____
_____
_____
_____

Managing calves after weaning:
_____
_____
_____
_____
_____

Name_____ Date_____

CHAPTER 15 MATCHING ACTIVITY

Term **Definition**

____ 1. integrated resource management (IRM)

____ 2. feed composition table

____ 3. rotation grazing

____ 4. carrying capacity

____ 5. husklage

____ 6. creep feeding

____ 7. artificial insemination

____ 8. castration

____ 9. preconditioning

____ 10. backgrounding

a. the number of animals that can be grazed on the pasture during the grazing season

b. gradually increasing the amount of grain provided for a calf as it grows

c. the growing and feeding of calves from weaning until they are ready to enter the feedlot

d. the removal of the testicles of the bull calves

e. a management tool available to beef producers that uses a team approach to improve the competitiveness, efficiency, and profitability of their beef business

f. the process of preparing calves for the stress of being moved into the feedlot

g. the analysis of feeds that can be used for rations for beef herds

h. the material that comes from the corn combine that can be used as a feed

i. the division of a field by the use of temporary fencing when feeding cattle

j. the placing of sperm in the female reproductive tract by other than natural means

Name_____ Date_____

CHAPTER 15 LAB QUESTIONS

1. Which specialists are needed in an IRM program?
   _____
   _____
   _____
   _____
   _____
   _____

2. What is an SPA, and how is it utilized by producers?
   _____
   _____
   _____
   _____
   _____
   _____

3. What are the major considerations when raising calves?
   _____
   _____
   _____
   _____
   _____
   _____

Chapter 16

Feeding and Management of Feeder Cattle

INTRODUCTION

Most cattle feeding operations exist in the grain-producing states of the north-central and southern parts of the United States. Beef cattle have high grain requirements for finishing; therefore, many cattle operations were established in or near the grain-producing states. The large commercial feedlots raising more than 1,000 head of cattle are found mostly in the Plains states, such as Colorado, Arizona, Texas, and California. The smaller feedlots of less than 1,000 head are found mainly in the Corn Belt and other locations across the United States.

Feeders are cattle that will be raised based on their daily rate of weight gain. Producers of feeder cattle have a goal of feeding each animal enough to make them grow at an appropriate daily rate so that their cost is less than the return of selling the feeder cattle. Some feeders use more roughage in the ration under a system of deferred finishing. Yearling feeders may be fed crop residues before being put on a full feed to clean up otherwise unused feed. Steers have higher weight gains in a feedlot than heifers. Calves make more efficient daily weight gains than older cattle, but they take longer to reach market weight. Crossbred feeder cattle have proven to gain weight faster than the pure parent breeds. More roughages can be used when feeding lower grades of feeder cattle because they are not producing high rates of gain.

Feeder cattle are purchased by the producer rather than using a cow-calf operation. Most feeders come from cattle auctions or are sold directly from the farm or ranch where they were born. The highest level of feeder cattle sales takes place during the month of October. Many producers sell their feeders at this time because, at this point, the productive season of pasture is winding down, the weather is getting colder, and the rate of gain will decrease while the cost of feed will increase. Using accurate records, a farmer can calculate the profit from feeding cattle. The profit is equal to the value of the finished cattle minus the total cost.

In the United States, most feeder cattle are fed corn; corn silage is the best roughage that can be used for increased rate of weight gain. For the protein requirements in a feeder cattle ration, many natural proteins are used, such as cotton seed meal, soybeans, and many others.

Other dietary requirements for feeders are salt, calcium, and phosphorus. It is sometimes necessary to add vitamins A, D, and E when finishing cattle.

One measure of feed efficiency is the feed intake of the cattle. Feed intake is influenced by the energy level in the ration, weather, feed palatability, feed processing, and the degree of finish on cattle. When starting cattle on feed, the proportion of roughage in the ration should be higher at the start and then gradually decreased as the animal grows and moves onto a full grain feed. The proportion of concentrate is gradually increased until the cattle are on full feed. The amount of roughages used before going onto full feed ranges from 10 to 70 percent; the proportion of feed not provided by roughages, will contain concentrates.

Because new cattle come on and off the farm often in a feeder cattle operation, the cattle must be watched closely for signs of sickness and disease. New cattle should be isolated in small lots with fresh feed and water. Vaccination and parasite control programs must be carefully planned and implemented soon after new cattle arrive. In this chapter, complete the feeder calculation exercise. It is good practice to plan the feeding regimen of various types and sizes of feeder cattle.

$$\text{Break-even feeder price} = \frac{\text{fed weight}}{\text{feeder weight}} \times \text{expected price of fed cattle} - \frac{\text{gained weight}}{\text{feeder weight}} \times \text{cost of gain}$$

Name_____ Date_____

EXERCISE 16-1 FEEDER CATTLE CALCULATION LAB

Beef cattle producers depend on every dollar per hundred weight of price for their animals. Usually, farmers will use a break-even analysis to measure their returns using the current market price and the current cost of production. Several cost vs. return calculators are available online for beef farmers to accurately estimate their profits or losses. The following website link provides, free to the public, beef cattle cost calculators. These calculators are made available by the University of Missouri Extension, Commercial Agriculture Beef Focus Team.

Feeder cattle calculator: Go to http://beef.missouri.edu/tools/ to access the Business Management and Marketing Tools page. Select the *Feeder Calf Evaluator* spreadsheet. Make sure to read the instructions before adjusting the calculator.

Directions: Refer to the following sample values for feeder cattle costs and returns. Input the sample values into the Feeder Calf Evaluator spreadsheet to determine (1) the return to ownership and management, (2) the purchase price to attain given return, and (3) the sale price needed to attain given return.

Determine the following performance measures using the values given in Table 16-1.

Cost of Interest

Average daily gain (deads out)

Average daily gain (deads in)

Total pounds of feed, as fed

Feed efficiency (lb of feed/1 lb of gain—deads accounted for)

Cost of gain (without return to management—deads accounted for)

1. Return to ownership and management

   _____
   _____
   _____

2. Appropriate purchase price

   _____
   _____
   _____

3. Sale price to attain given return

   _____
   _____
   _____

Name_____ Date_____

Table 16-1 Sample Feeder Cattle Costs

Price paid per pound ($)	$0.8600
Purchase weight (lb)	600.00
Trucking fee—arrival ($)	$3.50
Hired labor cost per head ($)	$2.00
Annual interest rate (that is 10)	10.00%
Price paid per pound of diet, as fed ($)	$0.050
Pounds of diet fed daily, as fed	10.00
Days owned	222.00
Yardage per day ($)	$0.29
Cost for processing products ($ for vaccines, implants, dewormers, tags)	$6.50
Morbidity rate (%)	5.00%
Cost of treatment ($) (entire cost of multiple-day treatment)	$12.00
Re-pull rate (%)	11.00%
Cost of re-pull O34 treatment ($)	$16.00
Mortality rate (%)	1.50%
Chronic rate	1.00%
Price received at sale ($/lb)	$0.9800
Unshrunk sale weight (lb)	1264.53
Percent shrink (%)	3.00%
Sales or commission fees ($)	$1.00
Trucking fee—departure ($)	$4.00
Decreased return for chronics (compared to pen average)	$130.00

Name_____ Date_____

CHAPTER 16 MATCHING ACTIVITY

Term

____ 1. efficiency in gain
____ 2. haylage
____ 3. boot stage
____ 4. high-moisture storage
____ 5. rolling
____ 6. crimping
____ 7. pelleting
____ 8. middling
____ 9. NEg
____ 10. bacterin

Definition

a. refers to harvesting the grain at a high moisture content (22 to 30 percent for corn and milo) and storing it in a silo

b. the same as rolling, except the rollers used have corrugated surfaces

c. refers to grinding the feed into small particles and then forming them into a small, hard form called a pellet

d. net energy for gain on a dry matter basis

e. refers to the amount of feed needed for each pound (kilogram) of gain

f. a liquid containing dead or weakened bacteria that causes the animal to build up antibodies that fight the disease caused by the bacteria

g. that point in growth at which the inflorescence (the flowering part of the plant) expands the boot

h. refers to processing the grain through a set of smooth rollers that are set close together; the grain is pressed into the form of a flake

i. low-moisture grass silage

j. a by-product of the flour milling industry; provides protein and energy in the ration and contains a good supply of phosphorus and potassium

Name_____ Date_____

CHAPTER 16 LAB QUESTIONS

1. Where are most feeder cattle operation located? Why?

   _____
   _____
   _____
   _____
   _____
   _____

2. What time of year are the most feeder cattle sold? Why at this time of the year?

   _____
   _____
   _____
   _____
   _____
   _____

3. What factors may affect feed intake?

   _____
   _____
   _____
   _____
   _____
   _____

Chapter 17

Diseases and Parasites of Beef Cattle

INTRODUCTION

When cattle are raised in confinement such as a beef lot, the presence of various diseases, bacteria, and parasites makes herd health management an integral part of the operation. Implementing a good herd health management program is essential because when cattle get sick, the profitability of those animals decreases. With the assistance of a veterinarian, a farmer should create a herd health prevention plan that will include several strategies to reduce the risk of a disease or parasite spreading. These strategies may focus on changing environmental factors, conducting routine herd testing, separation of groups of cattle, and even culling infected animals.

Vaccination is a tool routinely used by beef producers to reduce the threat of disease. Antibiotics and sulfa drugs are used to treat a multitude of diseases and infection. The concern is that large-scale use of certain antibiotics may lead to strains of the disease that are resistant to the medicine. When farmers buys replacement cattle, they should acquire cattle from disease-free herds and always isolate new animals for a period of time to help in disease control. Many diseases display similar symptoms, which can result in misdiagnosis. Laboratory tests and the aid of a veterinarian are often needed to accurately identify the disease that is present.

Insects spread many diseases, and their control is just as important in preventing the spread of disease in cattle. The most common external parasites of cattle are flies, lice, mites, and ticks. Some cattle operations use insecticide to control most of these parasites. All insecticide material must be labeled with directions that need to be followed carefully for safe use. Roundworms, flatworms, coccidia, and anaplasma are the most common internal parasites of beef cattle. When treating for internal parasites, sanitation is the most effective control.

Beef cattle are at risk to contract many health problems, and several modes of infection are possible. Vaccination, sanitary environment, and proper management of feeding programs help prevent some of these problems. Exercise 17-1 provides practice in identifying symptoms and implementing proper prevention and treatment programs.

Name_____ Date_____

EXERCISE 17-1 DISEASE PREVENTION AND TREATMENT ACTIVITY

Directions: For each of the following diseases or sicknesses that can affect beef cattle, identify the source of the disease, describe the symptoms that are displayed, and recommend common prevention and treatment methods.

Disease/Sickness	Sources of Disease	Symptoms	Prevention/Treatment
Bovine spongiform encephalopathy (BSE)			
Bovine virus diarrhea (BVD)			
Brucellosis			
Blackleg			
Foot and mouth disease			
Johne's disease (paratuberculosis)			
Leptospirosis			
Pinkeye (infectious keratitis)			
Shipping fever (pasteurella)			
Campylobacteriosis			

Name_____ Date_____

CHAPTER 17 LAB QUESTIONS

1. Which disease(s) may not show many symptoms until some of the cattle die suddenly?

   _____
   _____
   _____
   _____

2. What factors should a herd health plan address?

   _____
   _____
   _____
   _____

3. When buying replacement cattle, what steps should a producer take to reduce the spread of disease?

   _____
   _____
   _____
   _____

4. List three diseases that have one similar symptom.

   _____
   _____
   _____
   _____

5. Describe three ways to reduce the amount of insects around cattle.

   _____
   _____
   _____
   _____

Chapter 18
Beef Cattle Housing and Equipment

INTRODUCTION

When building a new cattle operation or expanding an established farm, detailed planning is necessary to avoid costly mistakes or setbacks. Well-planned facilities save labor and make the operation more efficient in many ways. When designing a facility, the producer should address issues such as employee and animal safety, allow for efficient animal handling, and reduce sources of stress for the animal. When planning a new facility, always plan for possible expansion in the future as the business grows. In general, cow-calf herds have the simplest facility requirements. The cows can calve on pasture, and in the winter months, some farmers will provide open-faced barns for shelter. Feedlot operations can range from widely complex to extremely simple, depending on the size and scope. Many feedlots use some form of automation during production to decrease labor costs.

When building confinement housing, warm barns do not make sense for beef operation because it does not increase the rate of gain. For beef cattle, cold confinement barns provide necessary shelter from the elements. Many of these confinement buildings are being built with slotted floors to make manure handling more efficient and to reduce the amount of manure in the barn. When designing a feedlot, windbreaks should be constructed to protect cattle from the weather.

Properly designed corral systems save cattle farmers a lot of time, reduce stress on the animal, and reduce injuries to cattle and humans. Researchers have recently made advances in understanding the factors that increase stress in corral systems. Corrals provide a place to sort and work with the cattle for herd checks and when treatments are needed; they also make it easier and safer to handle cattle. A corral should include holding pens, sorting pens, alleyways, a working chute, a squeeze chute, a headgate, and a loading chute. A cattle farmer may want to use a scale for accurate measurements of rate of gain and to determine the "on the hoof" weights of cattle being sold.

Feeding and watering facilities should be planned carefully as well. Feed storage facilities need to be designed; feed is stored in silos and feed bins. Feed bunks of various kinds are used to provide a controlled amount of feed to the cattle. Portable bunks are the least expensive and most commonly used. High-tech, mechanical feed bunks save labor but require a greater initial investment in equipment. An adequate supply of clean water is extremely important for a beef cattle facility. Automatic waterers are commonly used to reduce labor costs. Waterers need to be routinely cleaned to maintain sanitation. Other equipment used in beef operations includes mineral feeders, back rubbers, sunshades, creep feeders, and many other types of equipment to improve efficiency.

Name_____ Date_____

EXERCISE 18-1 DESIGN A MODERN CATTLE-HANDLING FACILITY LAB

When designing cattle-handling facilities, careful attention must be given to make the facilities effective to move, sort, and handle cattle safely. A curved corral design is usually beneficial when gathering and sorting beef cattle. There are several common designs for cattle corrals to avoid injury to the animals and to the cattle handlers. Figures 18-1 through 18-3 give examples of common facilities for reference.

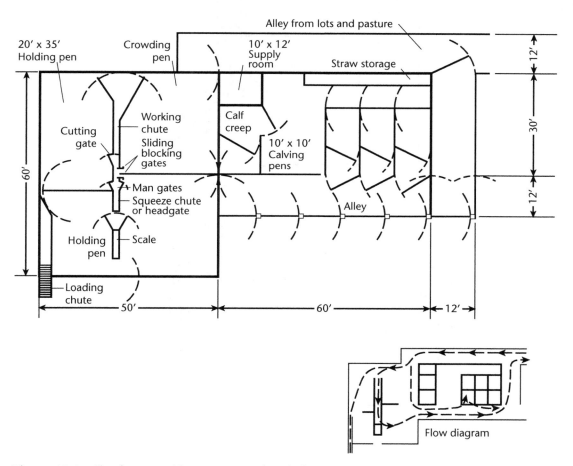

Figure 18-1 Plan for a corral for 300 to 1,000 head of cattle.

Name_____ Date_____

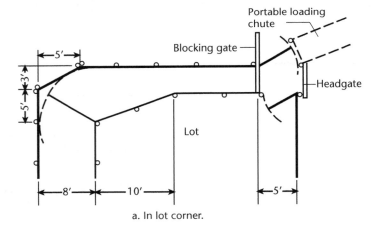

Figure 18-2 Two plans for small herds up to about 75 head.

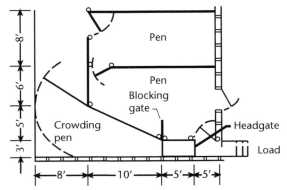

Figure 18-3 Half-circle chute. The entrance into a curved corral.

Name_____ Date_____

Directions: Create a design drawing for a cattle corral and handling chute. The drawing should be detailed and contain straight, clean lines. The drawing should be thoroughly labeled and include all dimensions. Use the working facility dimensions in Table 18-1 to make sure the design is appropriate for 600- to 1200-lb beef cattle.

Table 18-1

	Corral and Working Facility Dimensions		
	To 600 lb	**600–1,200**	**<1,200 and Cow-Calf**
Pen Space (sq ft/head)	14	17	20
Crowding Tub (sq ft/head)	6	10	12
Working Chute-vertical sides			
Width (inches)	18	20–24	26–30
Minimum Length (feet)	20	20	20
Working Chute-sloping sides			
Width at Bottom (inches)	13	15	16
Width at Top (inches)	20	24	28
Minimum Length (feet)	20	20	20
Working Chute Fence			
Height—minimum	45	50	60
Depth of Posts—minimum	30	30	30
Corral Fence			
Height	60	60	60
Depth of Posts—minimum	30	30	30
Loading Chute			
Width (inches)	26	26	26–30
Length (minimum, feet)	12	12	12
Rise, in/ft	3½	3½	3½

Dimensions from *Corral and Working Facilities for Beef Cattle.* GPE-5002

Create the design drawing in the following space (use a separate piece of paper if needed):

Name_____ Date_____

EXERCISE 18-2 EQUIPMENT COST LAB

When designing a cattle operation, one of the major limiting factors is cost. Some operations begin with a large amount of capital and are able to purchase all types of handling equipment, but many young start-up farmers do not have that kind of money available. This exercise is a class competition. The goal of the competition is to create the most cost-efficient facilities for the cattle operation. The design must be low cost, but it must also provide all necessary equipment for cattle to grow comfortably and facilities to handle the cattle. The student(s) who designs the most cost-efficient cattle farm will win the challenge.

Directions: A list of available equipment is provided here. Design and sketch a complete beef cattle facility. Include all the necessary facilities and equipment to feed and grow cattle, as well as load and unload cattle for transport. Create a cost analysis of the facility and turn it in to the instructor along with the sketch of the facilities. The student with the lowest cost analysis while still maintaining a functional operation will win the competition.

Equipment Price List
- Heated confinement barn for 500 head—$400,000
- Cold confinement barn for 500 head—$250,000
- Pour slab and cement curbs for outdoor feedlot for 500 head—$120,000
- Feedlot fencing for 500 head—$105,000
- Foundation fill (topsoil and stone) brought in as an alternative to cement—$8,000
- Slotted floors—$20,000
- Sub-floor manure lagoon—$80,000
- Outdoor manure lagoon—$50,000
- Bunk silo—$75,000
- Upright silo—$100,000
- Grain bins for 500 head—$25,000
- Portable mixer/grinder—50,000
- Sunshade for open feedlot for 500 head—$20,000
- Open-face run-in shed (each holds 100 head)—$10,000
- Back rubbers (each one services 50 head of cattle)—$2,000
- Creep feeders (25 animals each)—$1,000
- Self-feeders (25 animals each)—$5,000
- Cement bunk feeders for 500 head—$15,000
- Calf shelters (50 animals each)—$5,000
- Cattle-guard gates—$1,500
- Cattle treatment stall—$4,000
- Windbreak for feedlot—$20,000
- Corral system for 500 head—$15,000

Name_____ Date_____

    Headlock gate—$2,000

    Headlock chute—$5,000

    Headlock chute with tilting table—$10,000

    Automatic manure scrapers for allies for 500 head—$30,000

    Surveillance camera system for management—$5,000

Directions: Give this page to the instructor for competition evaluation.

Sketch of cattle facilities in the following space:

Cost analysis of facilities:

Name_____ Date_____

CHAPTER 18 MATCHING ACTIVITY

Term

____ 1. headgate
____ 2. tilting table
____ 3. dipping vat
____ 4. cattle guard

Definition

a. a device for restraining the animal in a horizontal position while it is being treated

b. allows equipment to be driven into an area without the operator having to stop and open a gate

c. used to hold the cattle while certain treatments are performed

d. used to treat livestock for pest control

Name_____ Date_____

CHAPTER 18 LAB QUESTIONS

1. What factors should be considered when planning a cattle facility?
   _____
   _____
   _____
   _____
   _____
   _____
   _____

2. What type of operation has the lowest cost? Why?
   _____
   _____
   _____
   _____
   _____
   _____
   _____

3. Why is new technology not always the best option when building cattle facilities?
   _____
   _____
   _____
   _____
   _____
   _____
   _____

Chapter 19
Marketing Beef Cattle

INTRODUCTION

There are several options for conventional beef cattle producers to market or sell their product. There are also some alternative markets for beef. The price that is paid for beef cattle is controlled by supply and demand. Producers have no control over demand—that factor is consumer driven. Producers do have some control over the supply, but other factors such as extreme weather conditions and available feed affect the supply of cattle as well. Beef cattle prices vary with the season of the year. Long-term trends or cycles in beef cattle prices have proven to drive price also.

There are three common methods through which beef cattle are marketed: terminal markets, auction markets, and direct selling. Terminal markets are also called public stockyards; the facilities of the terminal market are owned by a stockyard company, and the cattle remain the property of the sellers and buyers. The company charges for the use of the facilities and for feed fed to the cattle while they are in the stockyard. The importance of terminal markets has been declining in recent years because there are several other more convenient options for producers. Auction markets are found in many local communities and are also called sale barns. The numbers of cattle marketed in auction markets have not changed much over the years. More cattle are being sold by direct marketing methods now than they were a few years ago. There are several forms of direct marketing. Contract sales are used to market cattle directly from feeder cattle producers to cattle feeders. Cattle can be bought and sold by a cattle dealer or to order buyers that buy feeder cattle on order for cattle feeder operations. Online and electronic marketing have advanced in recent years, and a buyer can view and bid on cattle on home computer systems.

Purebred producers and value-added producers commonly use other direct marketing options such as private sales. In these marketing situations, the reputation of the breeder and the methods used in production and advertising are important factors in the sale of these specialty cattle. The sale of beef cattle that are transported across state lines is regulated by the Packers and Stockyards Act. This act created rules for fair business practices and competition in the marketplace, and it is enforced by the U.S. Department of Agriculture (USDA). Cattle shrink in weight when being shipped to market. Careful handling and management can reduce shrinkage loss and carcass damage that may occur during shipping.

Many producers select a marketing method based on convenience, but price and costs of marketing are also important factors that need to be considered. During sales, cattle are divided into classes based on use, age, weight, and sex. Feeder and slaughter cattle are given a grade to describe their body condition and conformation. Quality grade is determined by the age of the animal and the amount of marbling in the carcass. The yield grade is determined by the amount of lean meat that can be cut from the carcass. The grade given will affect the price paid for the animal.

Figure 19-1 Beef carcass during processing.

Another marketing option for producers is to use the cattle futures market to reduce price risk. A price is contracted based on futures market predictions, and the buyer and seller are locked in at that price for the duration of the contract. If the market prices drop, the producer's price will remain the same, but the producer will not see added profits if the market price increases. A cattle producer must keep good records and have an understanding of the pricing system to successfully use the cattle futures market.

In this chapter, the exercises will explore beef marketing methods and alternative markets.

Name_____ Date_____

EXERCISE 19-1 MARKETING METHODS LAB

A producer must choose a market based on the best options for his or her operation. A choice of market depends on many factors, such as price, size of animal, convenience for producer, and reasons for sale.

Directions: For each of the following market descriptions, determine if it is a terminal market, an auction market, and direct selling.

1. The Angus association in New York is hosting an annual feeder cattle sale. The marketing costs include charges for yardage, feed, insurance, brand, and health inspection. Charges are based either on a percentage of the selling price or a fixed fee.

   Marketing method: _____

2. The sale is made on the range or farm where the feeder cattle are produced. The producer enters a contract with the party that is buying the feeder cattle.

   Marketing method: _____

3. The producer consigns cattle to a commission firm. The firm does not take title of the cattle, but takes a fee to sort animals into uniform lots for sale and carry out the sale.

   Marketing method: _____

4. A cattle buyer uses an Internet auction program to see cattle that will be for sale in an online auction. The buyer views the cattle from video footage available online, and then enters a bid on the cattle.

   Marketing method: _____

5. About 12 percent of the meat packer sources of slaughter cattle are purchased from this type of cattle market.

   Marketing method: _____

6. An order buyer is a broker that buys slaughter cattle for packing plants. The cattle are bought on the farm and shipped directly to the packing plant.

   Marketing method: _____

Name_____ Date_____

EXERCISE 19-2　ALTERNATIVE BEEF MARKETS ACTIVITY

Some beef marketing strategies have unique factors and demand a higher price than conventional market options. The following is an example of a small-scale beef producer with value-added factors in their beef herd.

Sample Beef Cattle Operation

This beef operation consists of a cow-calf operation that raises the male calves for feeder stock and heifers as replacement heifers to grow the herd. The herd is purebred Angus cattle from top-grade genetics. The herd is also grassfed to add market value. The producer wants to get the top dollar value for the cattle but also considers convenience an important factor.

Directions: As a hired marketing consultant, choose an alternative market for the sample beef cattle operation. Describe how the producer will make a deal and transport the cattle as well as how the product will get to the end customers. Give the details as to why this is the best option for this farm. Compare it with other alternatives.

Alternative Market Analysis: (describe the suggested marketing plan here)

154 SECTION 4 Beef Cattle

Name_____ Date_____

CHAPTER 19 MATCHING ACTIVITY

Term

____ 1. supply
____ 2. demand
____ 3. stockyards
____ 4. commission
____ 5. yardage
____ 6. auction markets
____ 7. fill back period
____ 8. slaughter calves
____ 9. bullock
____ 10. quality grade

Definition

a. a fee for a firm's services in marketing the cattle
b. usually cattle between 3 and 8 months of age that have had feed other than milk in their diet for a period of time
c. the amount of a product that buyers will purchase at a given time for a given price
d. a male, usually under 24 months of age, that may be castrated or uncastrated and does show some of the characteristics of a bull
e. option where the facilities of the terminal market are owned by a stockyard company
f. defined as the amount of a product that producers will offer for sale at a given price at a given time
g. based on the amount and distribution of finish on the animal
h. the fee charged for the use of stockyard facilities
i. time during which cattle are fed after the cattle reach market
j. where cattle are sold by public bidding, with the animals going to the highest bidder

Name_____ Date_____

CHAPTER 19 LAB QUESTIONS

1. What is the beef check-off program? Describe.

2. How does the pork supply affect cattle prices?

3. Describe the different grades a beef carcass can receive. How does this affect price?

4. What is the Packers and Stockyards Act? How is it enforced?

SECTION 5

Swine

Chapter 20	Breeds of Swine	158
Chapter 21	Selection and Judging of Swine	165
Chapter 22	Feeding and Management of Swine	173
Chapter 23	Diseases and Parasites of Swine	179
Chapter 24	Swine Housing and Equipment	186
Chapter 25	Marketing Swine	192

Chapter 20

Breeds of Swine

INTRODUCTION

The major part of the swine ration is grain; this is why many of the swine operation in the United States are located in states where feed grain is produced in large quantities. More than half of the hogs raised in the United States are raised in the midwestern states. Swine are efficient at converting feed into meat; therefore, they have been a popular choice of livestock. Some swine producers jump in the market and buy pigs when the sale price is high and then leave the business when the prices are in a low trend. This practice is less common recently because many producers have invested in modern technology, equipment, and facilities, thus making it more economical to continuously raise swine.

Hog farmers often tend to use crossbreeds because they find many advantages to raising good crossbred swine. Selection of breeds to use in crossbreeding programs is based on data measuring litter size, growth rate, feed efficiency, carcass length, leanness, and muscle. The market demand for carcass quality pushes producers to manage many factors, from breeding to environment factors.

Of the many swine breeds in existence in America, several also have their origins in the United States. Other breeds were developed in Denmark and England. Just as in other livestock industries, purebred swine breeds have associations that register only purebred animals. Standards for registration of the individual breeds are established by the breed associations. Purebred swine are important for crossbreeding as well. Without the purebred breeding stock, the crossbred qualities would be diluted.

Controlled inbreeding has been developed and tested in the United States by state experiment stations as well as by private breeders. Inbred lines of hogs have been used in crossbreeding programs to improve litter size and growth rate. Another reproductive strategy becoming more relevant in swine production is genetic engineering. Genetic engineering holds promise of improving swine breed characteristics beyond what traditional breeding strategies have accomplished. This chapter develops a solid understanding of the common breeds of swine in the United States.

Name_____ Date_____

EXERCISE 20-1 BREED HISTORY AND FUNCTION ACTIVITY

In the modern swine industry, the performance and growth of pigs are measured more than ever. Only producers who intensively manage their herds are successful. Understanding the characteristics of different breeds is important knowledge for a pig farmer.

Directions: For each swine breed description, identify the matching pig breed. Answer all of the breed questions at the end of the exercise.

1. Breed: _____

   These black-bodied pigs have six white points, including their nose, tail, and feet. They have erect shorts ears and dished snouts. Legend says this breed was discovered by Oliver Cromwell's Army at Reading (the country seat of the shire of Berks) in England more than 300 years ago. They are known for providing hams and bacon of excellent flavor. They were first brought to America in 1823. Much improvement has been made through testing and genetic evaluation to meet the demand for fast, efficient growth, reproductive efficiency, and leanness.

2. Breed: _____

   Known as the durable mother breed, these pigs have white bodies with long, droopy ears. This breed originated in Chester County, Pennsylvania. More than 60,000 animals are recorded by this breeds' association each year. They have maintained their popularity with pork producers because of their mothering ability, durability, and soundness.

3. Breed: _____

   These red pigs with drooping ears are the second most recorded breed of swine in the United States and a major breed in many other countries. Their color can range from very light golden brown—almost yellow—to a very dark red that approaches mahogany. The growth of the breed is, in part, due to characteristics such as the ability to produce large litters, longevity in the female line, lean gain efficiently, carcass yield, and product quality as a terminal sire.

4. Breed: _____

   These black hogs have white belts across the shoulders, covering the front legs around the body. They have erect ears and are heavily muscled. They are the third most recorded swine breed in the United States. They are leaders in leanness and muscle, with good carcass quality, minimal amounts of back fat, and large loin eyes. Their ability to produce winning carcasses is unequaled, and they continue to set the standard by which all other terminal sires are evaluated. The females are known as great mothers and excellent pig raisers and have extra longevity in the sow herd.

5. Breed: _____

   These white pigs with large drooping ears are the fourth most recorded breed in the United States as well as a major breed in many other countries. Their purebred females are known for their ability to produce large litters over an extended time. Boars are aggressive and sire large litters that combine growth, leanness, and other desirable carcass traits. This, along with their outstanding maternal traits, have made them leaders in swine operations throughout the world.

6. Breed: _____

   These black-and-white–bodied pigs may have a white nose, tail, and feet. They have medium-sized, droopy ears and originated in the Miami Valley of Ohio in Butler and Warren Counties. This breed stands as the embodiment of perfection in the swine industry.

Name_____ Date_____

7. Breed: _____

These white-bodied pigs have black spots and medium-sized, droopy ears. Part of their ancestry can be traced back to the original Poland China hogs of Warren County, Ohio. This breed has continued to improve in feed efficiency, rate of gain, and carcass quality, as can be proven in testing stations throughout the country. They are popular with farmers and commercial swine producers for their ability to transmit fast-gaining, feed-efficient, meat qualities to their offspring.

8. Breed: _____

This white breed with erect ears is the most recorded breed of swine in the United States and Canada. They are muscular with a high proportion of lean meat and low back fat, in addition to being very sound. They are productive but are more performance oriented and durable than ever. The goal of the breed is to be a source of durable mother lines that can contribute to longevity and carcass merit. The motto "The Mother Breed and a Whole Lot More" indicates improvement and change in the industry.

Breed Bank:

| Berkshire | Chester White | Duroc | Hampshire | Tamworth |
| Landrace | Poland China | Spots | Yorkshire | Meishan |

Breed Questions

1. Which breed did you believe is the best? Why?

_____
_____
_____
_____
_____

2. How does knowledge about swine breeds help the efficiency of a pig farm?

_____
_____
_____
_____
_____

3. Why are some breeds more popular than others?

_____
_____
_____
_____
_____

© 2016 Cengage Learning®. May not be scanned, copied or duplicated, or posted to a publicly accessible website, in whole or in part.

Name_____ Date_____

4. List any other swine purebred breeds you can think of.
_____
_____
_____
_____
_____

Name_____ Date_____

CHAPTER 20 MATCHING ACTIVITY

Term

____ 1. swirl
____ 2. cryptorchidism
____ 3. freckles

Definition

a. when the skin has black pigmented spots
b. hair growing in a circular pattern from the roots; usually occurs along the top of the spine
c. males with one or both testicles retained

Name_____ Date_____

CHAPTER 20 LAB QUESTIONS

1. What are the five leading states in hog production? Which of them is number one?

_____
_____
_____
_____

2. Explain how crossbreeding combinations are selected in swine herds.

_____
_____
_____
_____

3. Which new technology has the ability to improve swine breed characteristics?

_____
_____
_____
_____

4. What is your opinion on the ethics of using this technology?

_____
_____
_____
_____

Chapter 21

Selection and Judging of Swine

INTRODUCTION

Most swine operations in the United States use crossbreeding to produce the desired traits of hogs. Terminal crosses are used to prevent the crossbreeding from becoming diluted or losing the traits that the producer wants in the hogs. Modern market hogs are more muscular, produce a larger loin eye area, and have less back fat than those produced only a few years ago. Because swine have relatively short gestation periods and large litters, changing the traits of market hogs does not take a long time.

Quality measurements and carcass grading have a strong influence over the price a hog will bring at market. Ultrasound is a recent technology that can measure the amount of fat-free lean pork in live hogs, helping producers to make decisions about which animals to keep for breeding stock. Pork packers can also use ultrasound measurements to help determine the value of the hog carcass. When swine are evaluated or judged, the evaluator mush have knowledge of the desired physical traits for the animal, a knowledge of the anatomy and terms, and the ability to read and review performance data. As a producer, it is necessary to be familiar with parts of the hog. Figure 21-1 in the textbook is a diagram of the pig anatomy with labeled parts.

The four primal cuts of the hog carcass are ham, loin, Boston shoulder (Boston butt), and picnic shoulder. These four cuts represent the most valuable part of the hog carcass. Hog producers try to control the quality of these primal cuts through breeding strategy and by using performance data to increase the probability of hogs with desirable traits.

Performance records are widely used by seed stock producers and market hog producers to evaluate breeding stock. When raising breeding stock, genetic control is an important part of the production program. Expected progeny difference (EPD) is a measure of how much of a change in performance on a given trait can be expected from the progeny (offspring) of a breeding animal when compared with the average of all animals in the population. EPDs can be compared within breeds and across herds but not across breeds. Sire summary and EPD values are available from several sources and can be easily accessed on the Internet. In addition to performance records, physical criteria are used when selecting breeding stock. These criteria include soundness of the animal and health of the breeding stock herd. When evaluating a group of swine, the following are important steps to follow:

1. Understand the class description, situation, or scenario.
2. Set priorities according to class description.
3. Evaluate the performance records.
4. Evaluate visual traits.
5. Decide on final rankings.

These steps can be used when evaluating a group of pigs at a judging contest, at a county fair, when buying a group of hogs, to measure success of a breeding program, or just in herd evaluation and measurement. This chapter provides examples to practice using performance data for evaluation of various groups of swine.

Name_____ Date_____

EXERCISE 21-1 USING PERFORMANCE DATA IN SWINE SELECTION

In this exercise, a list of tips and definitions are given. These will be helpful to evaluate two sample groups of swine using their performance data. The groups will need to be ranked in order of the pigs with the most desirable traits ahead of those with less desirable traits.

Tips, Facts, and Definitions

There are three categories of performance information: REPRODUCTION, GROWTH, and COMPOSITION. Performance records can be presented as (1) individual measurements, (2) ratios and/or indexes, and (3) genetic merit estimates (EPDs).

Category: Usefulness:	Individual Records Good	Ratios & Indexes Better	EPDs Best
REPRODUCTION	#farrowed	Dam's SPI	Born Alive EPD
	#weaned		
	21-day litter wt.	Dam's Index	SPI EPD
	Teat count		
	Dam's record		Maternal Index EPD
GROWTH	Average daily gain	Gain ratios	Days to 230 EPD
	Weight/day of age	NSIF Index	Terminal Index EPD
	Days to 230 lbs	F/G ratios	
	Feed efficiency (F/G)		
COMPOSITION	Backfat scan	BF ratio	BF EPD
	Loin muscle area		Terminal Index EPD

Figure 21-1 Categories of performance information.

Definitions

Sow productivity index (SPI): a measure of milking ability and prolificacy. The index combines number of pigs born alive and 21-day litter weight. It is adjusted for sow parity. The index is calculated with EPDs.

Terminal sire index (TSI): a measure of growth, efficiency, and backfat. The index is calculated with EPDs.

Maternal line index (MLI): places more emphasis on reproductive traits than growth traits. ; it is used when selecting replacement gilts. The index is calculated with EPDs.

BF: back-fat scan.

LEA: loin muscle area.

Name_____ Date_____

EPD number born alive (NBA): predicts the number born alive for daughters' litters relative to their farrowing group average. An individual with an EPD of +0.5 would be expected to produce daughters that would farrow litters with 0.5 more pigs than a sow with NBAEPD 0.0.

EPD litter weight: predicts the 21-day weight for their daughters' litters. An individual with an EPD of +3.5 would be expected to produce daughters that would wean litters 3.5 pounds heavier at 21 days than a sow with an EPD of 0.0.

EPD days to 230 pounds: predicts the growth performance of offspring in days to 230 pounds live weight. An individual with an EPD of −3.0 would be expected to produce progeny that reach 230 pounds 5 days faster than progeny of a parent with an EPD of +2.0.

EPD back fat: predicts the genetic contribution of potential parents for back fat of their progeny at 230 pounds live weight. For example, a boar with a back fat EPD (BFEPD) of −0.04 would be expected to sire progeny 0.04 inch leaner (10th rib BF) at 230 pounds than a boar with BFEPD of 0.00. Also, a gilt with BFEPD −0.05 would produce offspring 0.06 inch leaner than a gilt with BFEPD 0.01.

Directions: For the following examples of swine groups to be evaluated, use the performance data provided to rank the animals in order of most desired performance data to least desired performance data. For each animal, explain why they were ranked in that way using the performance information.

Situation 1: Hampshire Boars

The boars will be pasture bred to Yorkshire/Landrace sows. All offspring will be sold and marketed on a value-based marketing system. Make an initial ranking based on the given situation and the following production data.

The Hampshire boars in this class are going to be used as terminal sires. All offspring are going to be sold on a value-based marketing system. Therefore, the offspring need to be lean and heavily muscled to achieve high market premiums. No replacements are kept, so disregard the maternal data. Structural soundness is not a major priority in this class because the boars will be used on pasture versus confinement. The priorities for this class are leanness, muscle, and growth.

Boar #	Backfat	Days to 230 lb.	Loin Eye Area	# Born Alive	# Weaned	21-day Wt.
1	0.53	158	7.34	13	10	131
2	0.75	174	6.78	13	10	131
3	0.85	183	6.09	13	10	131
4	0.74	169	6.68	11	7	88

Figure 21-2 Performance data for sampled boars.

Name_____ Date_____

Rank these animals based on the situation and the performance data given for this group:

_____
_____
_____
_____
_____
_____

Explain your reasons for ranking the swine in this order:

_____
_____
_____
_____
_____
_____

Situation 2: Chester White Gilts

Rank the gilts as they should be kept in a seedstock herd that profits mainly from the sale of Chester White × Yorkshire F1 females to owners of terminal crossbreeding programs (terminal boars bred to Chester White × Yorkshire gilts). You and your customers raise all hogs in confinement. Your customers market all progeny on a value based system.

The Chester White gilts in this class are going to be bred to Yorkshire boars to produce F1 females that are going to be used in a terminal crossbreeding program. The F1 gilts will be bred to terminal sires. All progeny will be marketed on a value based system. Therefore, the F1 gilts must excel in maternal traits in order to produce large litters of fast growing market hogs.

Because the F1 female offspring will be expected to raise large litters, a high-quality underline should be emphasized along with the maternal figures. With the F1 gilts being placed in confinement, soundness is a priority. Even though the Chester White gilts will be used for maternal traits, they still must not cause a major setback in growth or leanness. The priorities for this class are maternal excellence, soundness, and high-quality underlines.

#	Days to 230lb EPD	10th Rib Fat EPD (in)	21-day Litter Wt. EPD (lb)	# Born Alive EPD
1	−2.9	+0.04	+0.5	+0.00
2	−2.4	+0.03	+3.7	+0.60
3	+0.8	−0.02	+1.7	+0.25
4	−2.6	+0.03	+3.7	+0.60

Figure 21-3 Performance data for sampled gilts.

170 SECTION 5 Swine

Name_____ Date_____

Rank these animals based on the situation and the performance data given for this group:

_____
_____
_____
_____
_____
_____

Explain reasons for ranking the swine in this order:

_____
_____
_____
_____
_____
_____

CHAPTER 21 MATCHING ACTIVITY

Term

___ 1. contemporary group
___ 2. estimated breeding values (EBV)
___ 3. expected progeny difference (EPD)
___ 4. porcine stress syndrome (PSS)
___ 5. sow productivity index (SPI)
___ 6. terminal sire index (TSI)
___ 7. maternal line index (MLI)

Definition

a. measure of growth, efficiency, and back fat
b. an inherited neuromuscular disease
c. a group of animals that are similar in a number of characteristics and have been raised under the same management practices
d. a measurement that places more emphasis on reproductive traits than growth traits
e. a selection index because it combines information from a number of sources to determine the genetic merit of the individual
f. a measure of milking ability and prolificacy
g. measures how much difference in performance on a given trait can be expected from the progeny of the breeding animal when compared with the average of all animals in the population

Name_____ Date_____

CHAPTER 21 LAB QUESTIONS

1. Why are quality measurements important in the swine industry?

_____
_____
_____
_____
_____

2. How is ultrasound used in the industry?

_____
_____
_____
_____
_____

3. What are the four most important primal cuts of a pig?

_____
_____
_____
_____
_____

4. What steps should be followed when evaluating a group of swine?

_____
_____
_____
_____
_____

Chapter 22
Feeding and Management of Swine

INTRODUCTION

Swine production requires intense management because quality factors affect the price directly. To remain competitive, swine producers must select breeding stock that will produce lean hogs and maintain feed efficiently. Management techniques such as split-sex and phase feeding can help increase production efficiency.

There are two main types of swine enterprises: those that produce purebred animals and commercial hog producers. Purebred animals only make up about 1 percent of the swine population in the United States. They are primarily raised for seedstock genetic improvement; it is a specialized industry, and the purebred producer must have thorough breeding management practices. The commercial industry has various types of operations such as feeder pig production, buying and finishing feeder pigs, and complete sow and litter systems (farrow-to-finish).

There are many types of feeds and sources of energy and protein that are used in the swine industry, but the two most common are corn and soybean meal oil. Corn has a high feeding value and soybean has a high level of available protein. Almost all commercial swine are produced in confinement housing for all phases of production, but there are some smaller operations that use pasture management to supply feed and protein while raising hogs. Hogs on pasture have a slower rate of weight gain than confinement hogs, which is a major reason it is not a large portion of the industry.

Whether fed a grain-based diet or pasture raised, hog have requirements for minerals, vitamins, and a clean, adequate source of water. Salt, ground limestone, and dicalcium phosphate are common sources of are sodium, chlorine, calcium, and phosphorus. These minerals are important for many functions of growth and reproduction. Trace minerals and vitamins are added to swine rations by using commercial feeds and mineral and vitamin premixes. Clean, fresh water needs to be readily available at free choice to the animal. Feeding rations differ for each stage of swine development, and a producer must manage the feeding program effectively.

Management of a sow through gestation and farrowing is labor intensive and a fragile time for both the sow and the litter of piglets. There are many strategies that are used to safely manage the sow and raise the piglets. A producer can save more pigs by being present when sows are farrowing, making sure each piglet nurses to receive colostrum, and making sure they are warm and dry. There are several necessary procedures that should be carried out to prepare the piglet for growth. Because the piglets' neurological sensors are underdeveloped, they cannot feel as much pain, and they recover quickly. A producer should clip needle teeth, ear-mark pigs, disinfect and clip the navel cord, and dock tails during the first day or two after farrowing. Iron shots should also be administered during the first few days after birth. Male pigs that will not be used for breeding should be castrated at 2 weeks old. A disease prevention program should be in place for piglets before they are weaned and enter groups with other pigs the same size.

The two exercises in this chapter focus on the various feeds that van be fed to sows during gestation and the volume needed, as well as farrowing practices once gestation has finished.

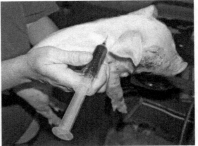

Name_____ Date_____

EXERCISE 22-1 FEED IMPACTS ON GESTATION LAB

Most operations that breed sows and raise them through gestation pay special attention to ensure that the sows get enough energy from feed to properly develop through the stages of pregnancy. Each producer has different available feeds depending on the region and the price of various commodities used in swine feed. In this exercise, there are several situations describing the type of feed available as the base feed for the ration as well as the metabolizable energy needed for the sow. Use the tables in the textbook for reference when developing the feed ration.

Directions: For each situation, determine how much of the base commodity is needed in the ration as well as all other ingredients needed to achieve the desired metabolizable energy levels. Record these values in the feed ration section of each situation.

Situation 1

The producer has good source of oats, soybean, and corn. The desired metabolizable energy for the operation is 1441 kcal/lb.

Feed ration: _____

_____

_____

Situation 2

The producer has a small supply of corn, soybean, and wheat middlings. The desired metabolizable energy for the operation is 1338 kcal/lb.

Feed ration: _____

_____

_____

Situation 3

The producer has a large supply of hard winter wheat, some soybean, and dehydrated alfalfa meal. The desired metabolizable energy for the operation is 1352 kcal/lb.

Feed ration: _____

_____

_____

176 SECTION 5 Swine

Name_____ Date_____

EXERCISE 22-2 FARROWING PRACTICES ACTIVITY

In this exercise, assume the role of the farrowing manager at a large farrow-to-finish operation. Being an operation with up to 10 sows farrowing a day, there are several employees helping to care for the sows and the piglets. The manager is responsible for successful farrowing, so a standardized operating procedure (SOP) sign is posted in the farrowing building, and an instruction binder is left in the office that outlines farrowing procedures.

Directions: Develop a standardized operating procedure. It should outline all of the considerations and procedures that need to take place during farrowing through weaning. This sheet should explain to the employee what the procedures are for care and management of the sow and piglets and why these steps are necessary.

SOP for Farrowing

_____
_____
_____
_____
_____
_____
_____
_____

Name_____ Date_____

CHAPTER 22 MATCHING ACTIVITY

Term

____ 1. PQA Plus
____ 2. scab
____ 3. fagopyrin
____ 4. ergot
____ 5. binding quality
____ 6. gossypol
____ 7. arsenicals
____ 8. flushing
____ 9. spray dried porcine plasma (SDPP)
____ 10. phase feeding

Definition

a. contained in cottonseed meal; is toxic to hogs
b. a by-product of blood from pork slaughter plants
c. a disease that attacks barley crops
d. designed to meet the rapidly changing nutritional needs of pigs during the first weeks after early weaning
e. a fungus that can infest rye and triticale
f. a management education program with major emphasis on swine herd health and animal well-being
g. means to increase the amount of feed fed for a short period of time
h. refers to how well the feed particles stick together in the pellet
i. a feed additive that increases efficiency in hog production
j. a photosensitizing agent that can cause rashes and itching when white pigs are exposed to sunlight

Name_____ Date_____

CHAPTER 22 | LAB QUESTIONS

1. Why are purebred swine important to the rest of the industry?

   _____
   _____
   _____
   _____

2. What are the two most common sources of energy and protein in the hog industry?

   _____
   _____
   _____
   _____

3. Why are these two sources used so often?

   _____
   _____
   _____
   _____

4. How old should a gilt be when it is first bred?

   _____
   _____
   _____
   _____

5. What is anestrus?

   _____
   _____
   _____
   _____

Chapter 23

Diseases and Parasites of Swine

INTRODUCTION

Swine production is a challenging industry because hogs are susceptible to many different diseases and parasites. Disease and parasite problems cause millions of dollars in losses each year in the U.S. swine industry. Prevention of health problems is the key to reducing losses. The major factors in maintaining healthy swine are sanitation, selection of healthy breeding stock, and proper management.

There are numerous infectious diseases that can affect swine, so a producer must be knowledgeable of prevention strategies and treatment procedures. Younger pigs are the most susceptible to disease. Vaccinations, sulfa drugs, and antibiotics are useful in the prevention and treatment of many of these diseases. The use of specific pathogen-free (SPF) hogs that have been born and raised in a sterile environment is also helpful in controlling some diseases.

Observing the vital signs in an animal can help in the early detection of health problems. Vital signs will vary according to activity and environmental conditions, but normal vital signs in swine are (1) temperature of 102.0 to 103.6°F, (2) pulse rate of 60 to 80 heartbeats per minute, and (3) respiration rate of 8 to 13 breaths per minute.

Some health problems arise because of feeding factors. Proper feeding levels and maintaining adequate body condition helps to reduce disease. The inclusion of minerals and vitamins is essential in preventing many diseases.

External parasites such as lice and mites can cause many health issues for hogs. A common condition is mange, which is caused by a tiny white or yellow mite that bores into the skin. The skin becomes irritated and results in the animal rubbing the affected areas. Mange is most severe in the fall, winter, and spring, and it spreads rapidly from hog to hog. Insecticides can be used to control mange, but all safety instructions and withdrawal periods must be followed.

Internal parasites can be very destructive internally and lead to severe sickness and low rates of gain in swine. The most serious internal parasite of hogs is the large roundworm. Hogs that are infested with internal parasites do not grow or gain weight as fast as other hogs, and they are more likely to have other health problems. Signs of a serious infestation of internal parasites are thinness, rough hair coat, weakness, and diarrhea. Some other worms that are major problems are lungworms, nodular worms, whipworms, and intestinal threadworms. A strict program of sanitation and treatment with drugs is necessary to control internal parasites.

Disease organisms may be carried into hog production areas by visitors. It is important to restrict the number of visitors allowed in and out of hog production facilities. Any visitor entering the facility should be wearing disinfected footwear. A container of disinfectant and a brush for disinfecting shoes should be kept at the entrance for the use of those entering the facilities. There are many steps that can be implemented with a disease prevention program.

The exercise in this chapter will require the exploration of each swine disease in detail. Understanding the causes, signs and treatment of the diseases is essential for developing a herd health program.

Name_____ Date_____

EXERCISE 23-1 DISEASES THAT AFFECT SWINE

Pigs are susceptible to more diseases than most species of livestock. A producer must understand all possible diseases and implement a prevention plan. When a disease has been contracted, a producer must be able to identify the signs and begin treatment and isolation of infected animals. As a swine producer, explain the protocol for disease identification and control.

Directions: For the diseases listed here, determine (1) the causes of the disease, (2) the signs that a pig has contracted the disease, (3) the effect the disease has on health and production, and (4) the treatment and prevention methods that should be used.

Abscesses

1. _____
2. _____
3. _____
4. _____

Actinobacillus pleuropneumoniae

1. _____
2. _____
3. _____
4. _____

Atrophic rhinitis

1. _____
2. _____
3. _____
4. _____

SECTION 5 Swine

Name_____ Date_____

Avian tuberculosis

1. _____

2. _____

3. _____

4. _____

Brucellosis

1. _____

2. _____

3. _____

4. _____

Cholera

1. _____

2. _____

3. _____

4. _____

Clostridial diarrhea

1. _____

2. _____

3. _____

4. _____

Name_____ Date_____

Swine dysentery

1. _____
2. _____
3. _____
4. _____

Edema

1. _____
2. _____
3. _____
4. _____

Eperythrozoonosis

1. _____
2. _____
3. _____
4. _____

Exudative epidermitis (greasy pig disease)

1. _____
2. _____
3. _____
4. _____

CHAPTER 23 MATCHING ACTIVITY

Term

____ 1. abscess

____ 2. specific pathogen free (SPF)

____ 3. rotavirus

____ 4. ampicillin

____ 5. turbinate bones

Definition

a. swellings filled with pus; the result of a bacterial infection that enters the body through the nose or mouth

b. located in the snout of the pig; they filter and warm the air before it reaches the lungs

c. antibiotic injection that helps prevent death

d. hogs that come from breeding stock that are surgically removed generally by cesarean section from the sows under antiseptic conditions; these hogs are raised in disease-free conditions

e. has similar signs as clostridial diarrhea

Name_____ Date_____

CHAPTER 23 LAB QUESTIONS

1. What are the normal vital signs for pigs?

_____
_____
_____
_____
_____

2. How is knowledge of the vital signs important?

_____
_____
_____
_____
_____

3. Give an example of an internal parasite. How does it affect the hog? How is it treated?

_____
_____
_____
_____
_____

4. Give an example of an external parasite. How does it affect the hog? How is it treated?

_____
_____
_____
_____

Chapter 24
Swine Housing and Equipment

INTRODUCTION

Pigs can be raised on pasture, but the majority of swine raised in the United States are raised in confinement housing. Housing is needed for farrowing, growing, and finishing hogs. A small number of producers still use pasture for all or part of the hog production cycle. These producers usually sell their pork as value added.

Swine operations have to be very strategic in their location for many reasons. The location of the farm in relation to neighbors and others is important because there are often complaints of the smell and traffic of daily operation at a hog farm. The direction of prevailing winds and the location of residences in the area must be considered when designing and building a swine facility.

Many producers in the swine industry are investing large amounts of capital into buildings and equipment in order to efficiently produce pork products. Modern confinement housing is equipped with proper ventilation to control temperature and moisture. Farrowing houses may be equipped with a heating source depending on the climate where the farm is located. Slotted floors are common to hog farms. They are an added cost, but they increase efficiency in manure removal and help keep hog pens clean. The plan for a hog enterprise must include fencing, handling equipment, feeding and watering equipment, crop production equipment, and any other automated equipment used on the farm.

Some farms with intensive management systems have higher equipment costs. There are many options for facility use from sanitation rooms for employees equipped with showers and sanitation suits, as well as the use of video cameras to monitor the animals especially during farrowing. Exercise 24-1 will give an example of equipment used for one swine operation. Study the data and determine the total market value of the farm's equipment.

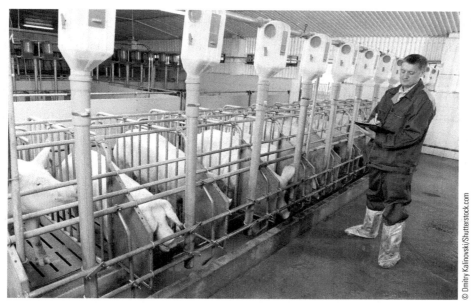

Figure 24-1 Example of modern swine housing.

Name_____ Date_____

EXERCISE 24-1 EQUIPMENT INVENTORY WORKSHEET

All equipment used by the farm or ranch should be included in the equipment resources inventory. This list should include any important information such as the model or serial numbers found on the equipment as well as the size, age, and condition of each piece of equipment. The inventory should also indicate whether the equipment is owned, leased, or borrowed. Finally, a current value should be included for each piece of equipment. Determining the value of equipment can be done by determining the book value of the piece of equipment and then deducting the depreciation from the value. The depreciation schedule will provide a value of the equipment that is equal to the original cost minus any accumulated depreciation.

Directions: Use the following example of equipment used on a small hog farm for equipment information and values. Enter the information into the Equipment Resources Inventory Worksheet to determine the total book value and market value (with depreciation) of all equipment on the farm.

Equipment Example

The operation has an older line of equipment and does not invest as much as some producers because their land and housing payments are quite high. The farm sees a depreciation value of 5 percent of the book value on all of its equipment. All of the equipment is owned outright by the farm with no financing loans. Most of the equipment was purchased when the farm started 5 years ago, unless stated otherwise. The equipment is all in good condition because the owners take care of their equipment. The operation owns a 200-horsepower tractor that is 5 years old and in good condition. It was bought for $79,000. There are 20-foot chisel plows and 15-foot offset plows worth $7,500 each. The farm has several grain drills to cover all of the land. All drills are 20 feet; one cost $12,000, one cost $13,000, and the third cost $9,000. There are two farm trucks: One is 5 years old and worth $10,000; the other is only 1 year old and worth $22,000. The farm has also various pieces of livestock equipment from farrowing crates to automatic waters, all valued at $8,000. Other miscellaneous ranch equipment is valued at $5,000 in total.

Name_____ Date_____

Equipment Resources Inventory Worksheet
List all Equipment Resources of the Operation
(Use Additional Pages if Necessary)

Equipment Name	Model #	Size	Purchase Year	Age	Condition			Ownership			Book Value	Market Value
					G	F	P	O	L	B		

Name_____ Date_____

CHAPTER 24 MATCHING ACTIVITY

Term

____ 1. warm confinement houses
____ 2. cold confinement houses
____ 3. growing period
____ 4. breeding rack
____ 5. zoning laws

Definition

a. used by counties, cities, and other local governments to define how property may be used
b. maintain desired temperatures regardless of the outside temperature
c. a ramp used for hand mating boars and sows
d. time from weaning to about 100 pounds
e. have temperatures that are only slightly warmer (3 to 10 degrees) than the outside temperature

Name_____ Date_____

CHAPTER 24 LAB QUESTIONS

1. What factors should be considered when planning a swine facility?

   _____
   _____
   _____
   _____
   _____
   _____

2. What factors affect the location for building a swine facility?

   _____
   _____
   _____
   _____
   _____
   _____

3. What equipment needs to be included in planning for a swine enterprise?

   _____
   _____
   _____
   _____
   _____
   _____

Chapter 25
Marketing Swine

INTRODUCTION

Marketing swine is an essential part of the business, and the decision of which market to use can make or break the profit margins for a hog producer. Markets vary on any given day in the price offered, the services given, and the costs charged to the seller and buyer. These factors change based on the number of hogs at the market that day as well as supply and demand. A producer must carefully study market trends to receive the best returns to the business.

There are four main avenues to market swine: direct marketing, auction markets, terminal markets, and group marketing systems. Most hogs in the United States are marketed through direct marketing; fewer producers market hogs through terminal and auction markets. The trend toward group marketing systems has been growing to help producers get higher prices for their pork.

Most swine in the United States is sold based on the weight of the animal. Producers that raise pork for quality standards get a better price by selling on yield and grade bases. When being marketed, swine are classified according to use, sex, weight, and quality. Butcher hogs and feeder pigs are graded according to USDA official grading standards. The grades of market hogs are U.S. No. 1, U.S. No. 2, U.S. No. 3, U.S. No. 4, and U.S. Utility, where No. 1 indicates the most desirable and Utility is used for the least desirable. This grading system is based on the quality of the lean meat and the percentage of the four lean cuts that the carcass will produce.

Currently, the best prices are typically paid for hogs weighing 220 to 260 lb (99.8 to 117.9 kg). Most producers aim to send hogs to market at 200 to 220 lb (90.7 to 99.8 kg). Large, meaty hogs can be raised over the common weight if the cost-to-price ratio remains positive. Hog prices tend to fluctuate with the seasons; they are generally lower in early spring and higher in the summer. Producers must account for the loss of weight due to shrinkage while the hogs are being transported through the marketing system.

The futures market may be used to reduce the risk of loss by hedging a fixed price. A producer using the futures market should understand the futures market and consult with experts.

Name_____ Date_____

EXERCISE 25-1 MARKETING METHODS LAB

A swine producer must choose a market based on the best options for its operation and the option giving the most value per pound of pork. A choice of market depends on many factors such as price, size of animal, convenience for producer, reasons for sale, and several herd considerations.

Directions: For each of the following market descriptions, determine if it is a terminal market, an auction market, group marketing, or direct selling. Be specific as to the method of marketing used.

1. A meat-packing corporation owns the local country buying station in an area of large swine production. The producer has a short distance to transport the swine, thereby decreasing shrinkage. The hogs are bought from the producer at a price based on the fat-free lean percentage of the carcass either on a percentage of the selling price or a fixed fee.

   Marketing method: _____

2. A large hog marketing cooperative is located in several states and increases the bargaining power of individual hog producers. This cooperative negotiates contracts with packers to supply large numbers of hogs from producers that are a part of the cooperative.

   Marketing method: _____

3. The producer consigns swine to a commission firm. There are usually several buyers competing for hogs, which may produce a better price. A commission fee will be taken out of the producer's check received for the hogs.

   Marketing method: _____

4. A hog farmer in a rural area raises slaughter hogs and then has them processed at a private USDA butcher. The producer pays a fee for processing, and all of the pork is returned processed, packaged, and labeled. The farmer then brings the pork to the city using a refrigerated truck and sells the product to restaurants and through farmers' markets.

   Marketing method: _____

5. A local livestock auction barn uses telephone hookups and video auctions to permit buyers who are not physically present to bid on the hogs.

   Marketing method: _____

Name_____ Date_____

CHAPTER 25 MATCHING ACTIVITY

Term

____ 1. boar taint
____ 2. skatole
____ 3. shrinkage
____ 4. futures contract
____ 5. hedging

Definition

a. the major cause of the odor problem with boar meat
b. when using the futures market, this is done to reduce the risk of loss if prices go down by locking in a price
c. odor given off by the meat of a boar
d. loss of weight as hogs are shipped to market
e. establishes a price for live hogs that are to be delivered at some future date

Name_____ Date_____

CHAPTER 25 LAB QUESTIONS

1. What is the pork check-off program? Describe.

_____
_____
_____
_____
_____
_____

2. How are the majority of hogs priced in the United States?

_____
_____
_____
_____
_____
_____

3. Describe the different grades a swine carcass can receive. How does this affect price?

_____
_____
_____
_____
_____
_____

4. Explain how the futures market works. Why would a hog producer choose to use the futures market?

_____
_____
_____
_____
_____
_____

SECTION 6
Sheep and Goats

Chapter 26	Breeds and Selection of Sheep	198
Chapter 27	Feeding, Management, and Housing of Sheep	204
Chapter 28	Breeds, Selection, Feeding, and Management of Goats	211
Chapter 29	Diseases and Parasites of Sheep and Goats	217
Chapter 30	Marketing Sheep, Goats, Wool, and Mohair	222

Chapter 26
Breeds and Selection of Sheep

INTRODUCTION

There are more than 40 breeds of sheep in the United States. Each breed has unique characteristics. Some sheep are large and some are smaller; some have black legs and heads, while others are white. Sheep are a major farm enterprise in the western range area of the United States, but in the rest of the country, farm flocks tend to be a secondary enterprise. In this case, cattle are usually the primary enterprise while sheep are mainly used to help maintain pasture.

Sheep are classified into several categories based on their wool production, meat production, breeding qualities, dairy production, and a combination of other functions. One way to classify sheep is by the type of wool they produce. Breeds of sheep that produce wool are separated into the following categories: fine wool, long wool, crossbred wool, medium wool, carpet wool, fur, and hair sheep. Fine wool breeding is most common in the western range areas, while the medium wool and crossbred wool breeds are more common in the farm flock area of the country. Hair sheep are becoming more popular in all parts of the United States as a low-maintenance meat breed.

As with most species of livestock, breed selection depends on the preference of the producer; therefore, sheep are divided into categories for evaluation. Depending on the function of the breed and the goals of the operation, different traits are valued when selecting sheep. The basic traits for sheep evaluation are overall health, soundness, fleece, conformation, and age. It is necessary to handle sheep when selecting or judging to determine traits such as muscling and finish. Market lambs are judged on the additional trait of finish which can be measured by touch.

The two exercises in the chapter will refine breed identification skills as well as market lamb evaluation skills. Knowledge and experience in breed and quality selection is important for all sheep producers.

Name_____ Date_____

EXERCISE 26-1 SHEEP BREED IDENTIFICATION

Sheep breeds can be grouped based upon their primary use in the sheep industry. Some sheep are raised to produce meat, and others are raised to produce wool. Some breeds are "ewe" breeds because they make good mothers and some are "ram" or "meat" breeds because they grow fast. Some sheep are dairy breeds and are raised to produce milk for making cheese. Sheep breeds can also be grouped based on the type of fiber they produce. Some sheep produce a fine, high-quality wool while other breeds produce medium- or coarse-fiber wool. Some breeds have hair and do not need shearing like the wool breeds. Some breeds are classed as "dual purpose" because they are acceptable for either meat or wool production.

Directions: For the following eight classifications of sheep, identify one example of a breed from each classification. Give a detailed description of the breed (history, functions, production characteristics, etc.).

1. Meat breed: _____
   _____
   _____
   _____

2. Fine wool breed:_____
   _____
   _____
   _____

3. Coarse wool breed:_____
   _____
   _____
   _____

4. Hair breed:_____
   _____
   _____
   _____

5. Ewe breed: _____
   _____
   _____
   _____

6. Black-faced breed: _____
   _____
   _____
   _____

Name_____ Date_____

7. Dairy breed: _____
_____
_____
_____

8. Dual-purpose breed: _____
_____
_____
_____

Name_____ Date_____

EXERCISE 26-2 MARKET LAMB LAB

Market lambs are judged on the basis of type, muscling, finish, carcass merit, yield, quality, balance, style, soundness, and smoothness. During this exercise, the instructor will provide a sample group of show-quality market lambs to be evaluated. A trip to a local sheep farm may be possible for a sample group, or a video of a judging class of market lambs may be used.

Directions: Use the following scorecard to describe and evaluate the sample market lambs. Record the information for 5 lambs, and then rank the lambs in order of most desirable traits.

Quality	Lamb 1	Lamb 2	Lamb 3	Lamb 4	Lamb 5
Type					
Muscling					
Finish					
Carcass Merit					
Yield					
Quality					
Balance					
Style					
Soundness					
Smoothness					
Rank Group					

CHAPTER 26 MATCHING ACTIVITY

Term

___ 1. banding instinct
___ 2. band
___ 3. staple
___ 4. NSIP
___ 5. western ewe
___ 6. native ewe
___ 7. type
___ 8. pelt

Definition

a. ewes produced anywhere in the United States other than the West
b. computerized, performance-based program for selection of sheep
c. consists of the skin and fleece
d. strong instinct to flock together
e. select body traits of sheep
f. a group or a flock
g. ewes that are produced in the western range area
h. refers to the fibers of wool

Name_____ Date_____

CHAPTER 26 LAB QUESTIONS

1. Describe one breed of fine wool sheep.
   _____
   _____
   _____
   _____
   _____

2. Describe one breed that is dual purpose.
   _____
   _____
   _____
   _____
   _____

3. What qualities are ewes evaluated for?
   _____
   _____
   _____
   _____
   _____

4. What qualities are market lambs evaluated for?
   _____
   _____
   _____
   _____
   _____

Chapter 27

Feeding, Management, and Housing of Sheep

INTRODUCTION

As in other livestock industries, sheep production can be divided into two categories: purebred and commercial production. There are four common types of sheep management systems based on the time of lambing: (1) fall lambs, (2) early spring lambs, (3) late spring lambs, and (4) accelerated lambing. Purebred producers produce breeding stock for commercial flocks and for other purebred flocks. The major goal of a purebred producer is breed improvement, so keeping good production records is necessary. Commercial producers maintain sheep flocks to produce meat and wool, so they use various breeds and crossbreeds to produce the desired traits.

When managing a flock of sheep, the feeding needs of the animals change as they enter different stages of development. The major stages of development that require specific feeding management are gestation feeding, lactation feeding, feeding the ram, flushing, feeding lambs to weaning, feeding lambs from weaning to market, feeding orphan lambs, and feeding replacement ewes. Each stage requires a custom ration with different amounts of roughage and concentrates. Pasture and good-quality legume or grass-legume hay is used as common roughage sources. Corn, grain sorghum, oats, barley, and wheat are grains are also commonly used in sheep rations. If good-quality pasture or hay is available, protein supplements are not needed. If necessary, soybean oil meal, cottonseed meal, linseed meal, and peanut meal are all good sources of protein to add to a sheep ration.

It is important that sheep producers do not allow breeding sheep to become too fat or too thin; they need a balanced body condition to effectively reproduce. Many breeds of sheep breed in the fall, but some sheep breed in other seasons as well. In order to have good breeding rates, a producer must prepare the ewe for breeding by clipping the wool from the vulva, udder, inside of the thigh, and around the eyes. The ram should be checked for fertility before the breeding season so there is time to find a replacement sire if needed.

Lamb care is very delicate, and it is easy to have lamb loss if proper management is not used. A producer can save lambs by making sure they nurse and do not become chilled. Depending on the location of the farm, warm confinement facilities may be needed. Lambs need to be docked, castrated, marked, and vaccinated to give them the best chance for survival. More concentrate than roughage is used to feed lambs to market weights.

In this chapter, the exercises provide situations requiring feed management and breeding decisions. Use the principles of sheep management described in the textbook to best complete the exercises.

EXERCISE 27-1 FEEDING CONSIDERATIONS

Maintaining feed with the proper available energy for the requirements of sheep at different stages of growth is imperative for efficient management. Producers try to avoid having sheep become too thin or gain too much condition. The appropriate feeding ration is based on the stage of development of each animal.

Directions: For each of the following sheep at different stages of life, describe the suggested feeding strategy for proper body condition maintenance. (Suggested feeding rations are provided in the textbook.)

1. **Gestation feeding**
   _____
   _____
   _____
   _____

2. **Lactation feeding**
   _____
   _____
   _____
   _____

3. **Feeding the ram**
   _____
   _____
   _____
   _____

4. **Flushing**
   _____
   _____
   _____
   _____

5. **Feeding lambs to weaning**
   _____
   _____
   _____
   _____

6. **Feeding lambs from weaning to market**
   _____
   _____
   _____
   _____

Name_____ Date_____

7. Feeding orphan lambs
_____
_____
_____
_____

8. Feeding replacement ewes
_____
_____
_____
_____

Name_____ Date_____

EXERCISE 27-2 CROSSBREEDING ACTIVITY

Many sheep producers use crossbreeding to grow ewe and ram lambs with desirable traits. Most farms use a three-breed cross for the best results. Although most breeds have similarities, there are traits that are more dominant in one breed over another. The breeds in a cross must be chosen carefully, and the producer needs to have a good understanding of the traits each breed provides. Use Chapter 26 in the textbook as a reference for the traits of each breed.

Directions: Below are five examples of crossbred lambs that possess specific genetic traits. For each example, identify three sheep breeds that can be crossbred to produce a lamb with those traits.

Lamb 1: Breed A_____ Breed B_____ Breed C_____

Description: The traits of this crossbreed include wrinkled skin and medium, angular bodies. They are white faced with wool on their legs. They do not have a strong flocking instinct.

Lamb 2: Breed A_____ Breed B_____ Breed C_____

Description: The ears, nose, face, and legs are white. This crossbreed produces a medium-coarse fleece that averages 4 to 5 lb. They are large in size and have a blocky body type.

Lamb 3: Breed A_____ Breed B_____ Breed C_____

Description: They have no wool on the face and legs. The face, ears, and legs are gray to brown, and the lamb crop is often 150 percent.

Lamb 4: Breed A_____ Breed B_____ Breed C_____

Description: They have gentle disposition and are economical to feed and maintain. The lambs grow rapidly and deposit fat at a slower rate than other breeds. The breed is polled, and the color of the face is reddish brown to bright tan.

Lamb 5: Breed A_____ Breed B_____ Breed C_____

Description: This sheep is polled, and it has a tuft of wool on the forehead. It produces 12 to 16 lb of fleece and is better adapted to wet, marshy areas than are other long wool breeds of sheep.

Name_____ Date_____

CHAPTER 27 MATCHING ACTIVITY

Term
____ 1. hot house lambs
____ 2. accelerated lambing
____ 3. drench
____ 4. tagging
____ 5. wether
____ 6. prolapsed uterus
____ 7. extension hurdle

Definition
a. occurs when the uterus protrudes from the vulva
b. refers to shearing around the udder, between the legs, and around the dock
c. a system that produces three lamb crops in 2 years
d. a male lamb that has been castrated before reaching sexual maturity
e. a portable gate that can be lengthened or shortened as needed to crowd sheep into a corner
f. lambs that are sold at 50 to 90 days of age; they weigh 35 to 60 lb
g. a large dose of medicine mixed with liquid and put down the throat of the animal

Name_____ Date_____

CHAPTER 27 LAB QUESTIONS

1. What are the main goals of a purebred sheep producer?

   _____
   _____
   _____
   _____
   _____

2. What procedures are essential to keep lambs alive after they are born?

   _____
   _____
   _____
   _____
   _____

3. What are the common ingredients used in sheep feeding rations?

   _____
   _____
   _____
   _____
   _____

Chapter 28

Breeds, Selection, Feeding, and Management of Goats

INTRODUCTION

There has been a national trend toward higher consumption of goat products, including meat, milk, and other products such as cashmere. At one time, goat products were only available in health-food stores or specialty shops, but now goat meat, milk, cheese, and butter can be found in local supermarkets across the country. This increase in demand has changed the scope of the goat industry from a small niche market to a large, popular market with many new producers growing goats.

One of the major reasons for the increased production of goats is the growth of the ethnic populations in the United States who view goat meat as customary in their diet. Other reasons for the increase in the production include the ability of goats to utilize land unsuitable for other livestock, an increase in demand for products due to heath concerns, the rising popularity of goat meat as delicacies in more upscale restaurants, and the increase in popularity as show animals.

The three major classifications of goats are dairy, fiber, and meat. There are breeds of goats that were also established for pets/companions and for goatskin. Small herds of dairy goats are common in all parts of the United States due to the increase in hobby farms.

The major breeds of dairy goats found in the United States are Alpine, LaMancha, Nigerian Dwarf, Nubian, Oberhasli, Saanen/Sable, and Toggenburg. Dairy goats produce much less milk than cattle, but they are still selected for similar traits and high milk production. High-producing dairy goats average 3 to 4 quarts of milk per day.

Although there are many breeds of fiber sheep, the three fiber goat breeds that are common in the United States are Angora, Cashmere, and Miniature Silky Fainting Goats. Fiber goats are evaluated on the basis of body type and fleece. Breeding animals are further selected on the basis of age and fertility.

The common meat type goat breeds in the United States are Boer, Kiko, Kinder, Myotonic, Pygmy, Savanna, and Spanish. When selecting goats for breeding stock, producers must consider the purpose, functionality, and durability of the animal.

Goats are ruminants and have digestive systems similar to those of cattle, so roughage may be used as large part of the ration when feeding goats. Producers will supplement grain when low-quality roughages are being fed in order to provide additional nutrients during growing and lactating periods.

Each goat operation will have different housing needs depending on the size and scope of the operation. More equipment and facilities are necessary for dairy goats. These facilities depend on the number of goats owned and the convenience desired by the producer. When running a dairy goat operation, an essential aspect of the business is keeping the milk and milking equipment clean.

This chapter explores a few of the numerous goat breeds in existence throughout the United States. Complete the exercises in order to develop knowledge and skill in breed identification and goat facility design.

Name_____ Date_____

EXERCISE 28-1 GOAT BREED IDENTIFICATION

There more than 90 breeds of goats available to producers depending on what desired traits suit the livestock operation. Knowledge of the breed descriptions is necessary when selecting a breed to produce.

Directions: For each number, use the description to correctly identify the breed name.

1. _ _ _ _ _ _
2. _ _ _ _ _
3. _ _ _ _ _
4. _ _ _ _ _ _
5. _ _ _ _ _ _
6. _ _ _ _ _
7. _ _ _ _ _
8. _ _ _ _ _ _

Descriptions

1. This breed originated in France, has upright ears, and can be any color or combination of colors. It has a straight face, has medium to short hair, and is medium to large in size.

2. This is the only breed developed in the United States. It has either "gopher" or "elf ears." Any color or combination of colors is acceptable, and it has short, fine, glossy hair.

3. A Swiss breed of rugged bone, this breed is medium to large in size and either white or cream in color. It has short and fine hair, is erect-eared, and has either a straight or dished face.

4. A Swiss breed known for upright ears, straight faces, and chamois color, it has a black belly and a light gray to black udder. One of the smaller Swiss breeds, it is a minimum of 28 in. in height and is the newest recognized breed by the ADGA.

5. This breed was originated in the Himalaya Mountains of Asia and has a straight or concave nose, pendulous ears, and twisted horns. It is usually a small, white breed with a long, fine, and lustrous mohair fiber coat. The fine underwool is a valuable product called cashmere. This breed is known primarily as a browsing animal.

6. This breed came from West and Central Africa and the Caribbean. Dwarf, short legged, hardy, and alert, its profile should have a dished appearance with a broad, strong, and well-muscled jaw. It has a small compact body, and its main colors are white caramel, caramel, gray agouti, black agouti, and charcoal.

7. This breed originated in India and Egypt and is known for its high quality, high butterfat concentration, and high milk productivity. It has a strong, convex facial profile between the ears and the muzzle and long, bell-shaped, wide ears. It can have any color pattern and has short, glossy, and fine hair.

8. Of Swiss origin, this breed is medium in size, has upright ears and a dished or straight face, and is solid-colored, varying from light fawn to dark chocolate. It has white ears with dark spots in the middle, two white stripes down the face from each eye to the muzzle, white hind legs, and a white triangle on either side of the tail. It is known for its high milk productivity.

Name_____ Date_____

EXERCISE 28-2 DESIGN A GOAT FACILITIES LAB

Designing a dairy goat facility takes many different factors into consideration. Some of these considerations include using loose pens or tie-stall barns, design of the building structure, the size of pens and room for each animal, providing water, providing feed, climate considerations, corrals and handling equipment, and various styles of milking facilities and equipment, as well as milk storage equipment. This exercise focuses on the design of a dairy facility that includes all of the necessary equipment.

Directions: The facility design below is an example of a basic dairy goat operation. Fill in the facility map with all of the equipment necessary to milk goats. The added equipment and facilities should be drawn on the map and labeled. Creating a key to the drawings of equipment may help keep the map neat and legible.

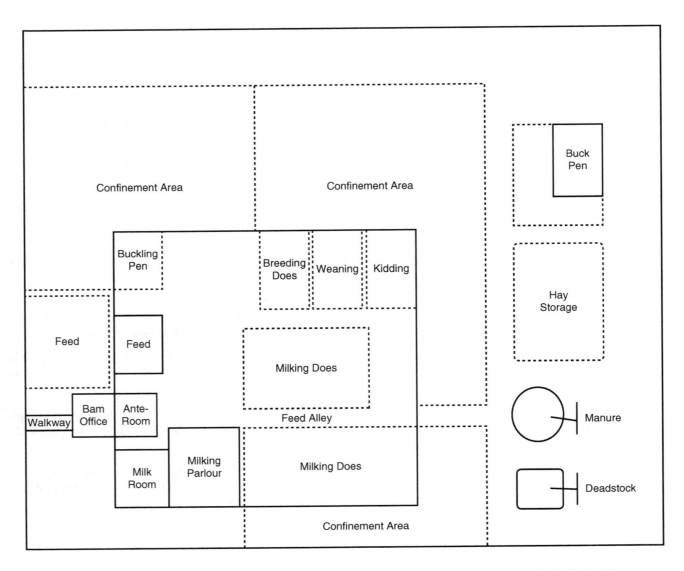

Name_____ Date_____

CHAPTER 28 MATCHING ACTIVITY

Term **Definition**

____ 1. buck   a. young goat under 1 year of age, either sex
____ 2. doe   b. refers to spasm or temporary rigidity of the muscles
____ 3. doeling   c. a small segment of a larger market
____ 4. kid   d. male goat, any age
____ 5. wether   e. the shoots, twigs, and leaves of brush plants
____ 6. niche market   f. a projection of skin hanging from the chin
____ 7. browse   g. female goat, any age
____ 8. wattles   h. refers to the soft down or winter undercoat of fiber produced by most breeds of goats
____ 9. cashmere   i. an unbred female goat
____ 10. mytonic   j. male goat castrated when young

Name_____ Date_____

CHAPTER 28 LAB QUESTIONS

1. What recent changes have occurred in the goat industry?

   _____
   _____
   _____
   _____
   _____

2. Which factors must be considered when selecting goats for breeding?

   _____
   _____
   _____
   _____
   _____

3. What is the other term used to describe mytonic goats?

   _____
   _____
   _____
   _____
   _____

Chapter 29

Diseases and Parasites of Sheep and Goats

INTRODUCTION

In order to maintain a good health program for sheep and goats, a producer must focus on prevention of infection rather than treatment. Most treatments are costly and not very effective. When sheep and goats become sick, they rarely recover because, when signs of illness are noticeable, the infection has usually progressed beyond treatment. Without a proper treatment plan, profits on sheep and goat farms decrease significantly. Good management, feeding, vaccination, and sanitation programs help prevent health problems. The involvement of a veterinarian is recommended when developing a treatment plan.

Sheep and goats have many diseases in common with cattle, but there are several that specifically affect these two species. These include blue tongue, enterotoxemia, foot rot, foot abscess, dysentery, pneumonia, scrapies, tetanus, and several others. The symptoms of many of these diseases are similar, so identifying the sickness may take laboratory tests and the assistance of a veterinarian. Many of these illnesses can be prevented using available vaccines. If an animal becomes infected, treatment with antibiotics, drenching, and isolating the sick animal are basic control measures.

The most common internal worm parasites of sheep and goats are stomach worms, intestinal worms, lungworms, and liver flukes. The eggs of most of these internal parasites are spread over pasture grasses through animal feces. Frequent pasture rotation and management are needed to prevent or reduce infections. Allowing pastures to rest between grazing periods gives time for larvae to die before they are consumed while animals are grazing.

Raising sheep and goats can be challenging because a large number of nutritional health problems can affect the animal. Many of these infections can be prevented or cured with good feeding management. Some health problems are related to nutritional factors, including night blindness, milk fever, impaction, pregnancy toxemia, and consuming poisonous plants.

The exercises in this chapter involve recommending tapeworm prevention strategies and identifying various parasites that affect sheep and goats.

Name_____ Date_____

EXERCISE 29-1 TAPEWORM LAB

The infection of worms presents a significant production threat to sheep, costing the industry millions of pounds a year. The main worms affecting sheep are roundworms, tapeworms (cestodes), and lungworms. Figure 29-1 shows the process that takes place when a sheep is infected with tapeworms. A producer must understand the cycle of parasite infection and change management strategies to reduce or eliminate the infection of parasites such as tapeworms.

Directions: Study the path that a tapeworm takes as in infects a sheep. Once the cycle is understood, suggest five management strategies that could help a sheep or goat farmer reduce or eliminate the presence of tapeworms in the animals.

Tapeworm Prevention Recommendation:

1. _____
2. _____
3. _____
4. _____
5. _____

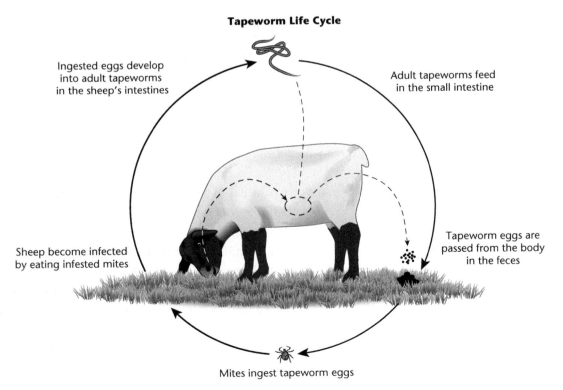

Figure 29-1 Life history of a tapeworm of sheep.

Name_____ Date_____

EXERCISE 29-2 PARASITE IDENTIFICATION LAB

Directions: For each of the following internal parasites of sheep and goats, (1) determine the scientific name, (2) the characteristics of the parasite, (3) the effect the parasite has on the animal, and (4) possible treatment or prevention strategies. Some information may need to be gathered using Internet sources.

Common Stomach Worm:

1. _____
2. _____
3. _____
4. _____

Bankrupt Worm:

1. _____
2. _____
3. _____
4. _____

Lung Worm:

1. _____
2. _____
3. _____
4. _____

Liver Fluke:

1. _____
2. _____
3. _____
4. _____

CHAPTER 29 MATCHING ACTIVITY

Term

___ 1. virus
___ 2. toxemia
___ 3. convulsions
___ 4. causative agent

Definition

a. involuntary muscle contractions

b. a self-reproducing agent that is considerably smaller than a bacterium and can multiply only within the living cells of a suitable host

c. a biological pathogen that causes a disease; generally bacterial, fungi, or viral

d. a condition resulting from the spread of bacterial toxins through the bloodstream

Name_____ Date_____

CHAPTER 29 LAB QUESTIONS

1. Why is treatment of sickness not usually successful with sheep and goats?

   _____
   _____
   _____
   _____
   _____

2. What is the most effective strategy to control infection of disease and parasites?

   _____
   _____
   _____
   _____
   _____

3. List the most common parasites in sheep and goats.

   _____
   _____
   _____
   _____
   _____

4. Describe the lifecycle of parasites and how they enter the bodies of the animals.

   _____
   _____
   _____
   _____
   _____

Chapter 30
Marketing Sheep, Goats, Wool, and Mohair

INTRODUCTION

The sheep and goat industries in the United States are small in comparison with other livestock industries such as swine, poultry, beef, and dairy. Although the numbers of sheep marketed in the United States have been decreasing, the industry is looking to reverse the trend through niche markets and a growth of small start-up operations.

The American Lamb Board and American Wool Trust provide national promotion of sheep and wool products through advertising, public relations, and retail promotions. There are several marketing methods available to sheep producers: terminal markets, local pools, sale barns, direct to the packer, or electronic marketing. The producer must choose the option that works best for the operation based on the kinds of markets available locally, prices paid, the numbers of lambs to be sold, and the transportation options.

Sheep are classed according to age, use, sex, and grade. Age classes are: lamb, yearling, and sheep. The use classes are: slaughter sheep, slaughter lambs, feeder sheep, feeder lambs, breeding sheep, and shearer lambs. The sex classes are divided into: ewes, rams, and wethers (castrated males). When marketing meat sheep, grades are given for live animals that are directly related to the quality and yield grades of the carcasses they will produce. Yield grades are based on: the measure of back fat, predicted yield of the carcass, quality grades on the body conformation, and amount of finish.

Wool is marketed through local buyers, wool pools, cooperatives, or warehouse operators or by direct sale to wool mills. Each producer will choose a wool market depending on several local factors, including number of fleeces to market, transportation requirements, price paid, and convenience. The value of the fleece is determined by the amount of clean wool that it produces, the length of the fibers, the density, and the diameter of the wool fibers (Figure 30-1).

Goat meat is called chevon and is mostly marketed directly to the customer. Young kids that weigh 30 to 40 lb are marketed as cabrito, which is considered a delicacy. Explore marketing opportunities of sheep through the exercises in this chapter.

Figure 30-1   Retail wool products.

Name_____ Date_____

EXERCISE 30-1 CLASSING AND GRADING OF SHEEP

Sheep are classified into four categories: age, use, sex, and grade. These factors change the value of the sheep carcass, the texture, and the taste. Live grades of sheep are based on quality and estimated yield. This exercise will provide information on two different sheep to bee classed and graded.

Directions: Based on the information provided in the following examples, determine (1) the class, (2) the quality grade, and (3) the yield grade for each sheep.

Sheep 1

This ram is almost 3 months old and weighs 55 lb. It is moderately wide and thick in relation to his length; it has moderately plump and full legs, a moderately wide and thick back, and moderately thick and full shoulders. The lean flesh and external finish are firm; the flanks are moderately full and firm. It has a back fat measurement of 0.16 to 0.25 in.

Class: _____

Quality grade: _____

Yield grade: _____

Sheep 2

This ewe is 9 months old. It is moderately narrow in relation to her length; has slightly thin, tapering legs; and has slightly narrow and thin back and shoulders. It has moderately firm, lean flesh and external finish and has slightly full and firm flanks. It has a back-fat measurement of 0.26 to 0.35 in.

Class: _____

Quality grade: _____

Yield grade: _____

Name_____ Date_____

EXERCISE 30-2 MARKETING WOOL

There are four common methods to market sheep wool. A producer must understand each market option and choose the best one according to convenience and the price received. Part 1 will require a knowledge of each marketing method, and Part 2 will examine sheep operations in the local area and determine which marketing method best suits the local wool business.

Part 1

Directions: Describe each of the following marketing options in detail. Explain who is involved, how money is exchanged, and what the producer can expect when it takes the wool to the market.

Wool pools: _____
_____
_____

Cooperatives: _____
_____
_____

Warehouse operators: _____
_____
_____

Direct sale to wool mills: _____
_____
_____

Part 2

Evaluate the local wool industry. What size sheep farms are in the county? Which wool markets already exist in a reasonable distance? Make a suggestion of an existing local market where producers can take their wool. Explain why this is the best marketing choice.

Suggested market for local producers: _____
_____
_____

Name_____ Date_____

CHAPTER 30   MATCHING ACTIVITY

Term

____ 1. breeding sheep
____ 2. shearer lambs
____ 3. cutability
____ 4. feathering
____ 5. grease
____ 6. spinning count
____ 7. wool top
____ 8. apparel wool
____ 9. worsted
____ 10. clip

Definition

a. the number of hanks of yarn that can be spun from 1 lb of wool top
b. process that uses longer fiber wools
c. intermingling of fat with lean
d. the wool or mohair produced by a single shearing
e. western ewes that are sold back to farms and ranches to be bred to produce more lambs
f. used in the making of clothing
g. refers to animals not finished enough for slaughter; they are shorn and fed to a higher level of finish before slaughter
h. the presence of impurities in the fleece
i. refers to the yield of closely trimmed, boneless retail cuts that come from the major wholesale cuts
j. partially processed wool

Name_____ Date_____

CHAPTER 30 LAB QUESTIONS

1. Which two associations provide promotion of sheep products?

   _____
   _____
   _____
   _____
   _____

2. What is the most common method of marketing chevon?

   _____
   _____
   _____
   _____
   _____

3. What qualities are wool grades based upon?

   _____
   _____
   _____
   _____
   _____

SECTION 7

Horses

Chapter 31	Selection of Horses	230
Chapter 32	Feeding, Management, Housing, and Tack	239
Chapter 33	Diseases and Parasites of Horses	247
Chapter 34	Training and Horsemanship	255

Chapter 31
Selection of Horses

INTRODUCTION

The use of horses in the United States has changed drastically over the past 150 years. Horses are no longer needed as a tool for agriculture, transportation, and military use. The horse population has decreased over this time, but horses have remained popular and have been adapted for use to fulfill many other purposes. Today, it is estimated that the total horse population of the United States is about 9.2 million. Seventy-five percent of the horses are used for personal recreation and pleasure; remaining 25 percent horses are used for ranching, racing, breeding, and commercial riding.

Although the population of horses has decreased, the horse industry is still big business. In the United States, the equine industry accounts for more than $39 billion of direct economic activity annually. Much of this economic activity is generated by the horse racing industry and by recreational uses of horses. The industry provides a large number of jobs and related careers. More than 4.5 million people are involved in the horse industry as owners, service providers, employees, or volunteers.

There are many breeds of horses, and each breed has been developed to perform specific tasks. Horses are classified as either a light breed, a pony breed, or a draft breed. Most of the horses in the United States are of the light breeds. Today, there are fewer breeds of draft horses in the United States due to the decreased use of draft horses in the agriculture industry. Ponies are smaller than light horses. Draft horses are larger than light horses.

Horses should be selected on the basis of conformation, uses performed, the age of the animal, sex, and soundness. Breed selection is a matter of function and personal preference. The five basic colors of horses are bay, black, brown, chestnut, and white. There are different variations of color, which include dun, gray, palomino, pinto, and roan. Horses may suffer from several unsoundnesses and blemishes. Unsoundnesses affect the physical performance of a horse. The most serious unsoundnesses affect the feet and legs of the horse. A blemish is visible, but does not affect the performance of a horse.

CHAPTER 31 Selection of Horses 231

Name_____ Date_____

EXERCISE 31-1 BREED SELECTION

The most popular breeds of horses in the United States, based on purebred registrations, are Quarter Horse, Thoroughbred, Paint, Appaloosa, and Arabian; they make up 75 percent of the total horse population. Many other breeds exist in small numbers. This exercise will test the ability to identify various breeds of horses.

Directions: For each of the following pictures of horses, identify that breed and list three breed characteristics (A, B, and C).

1.

A. _____
B. _____
C. _____

2.

A. _____
B. _____
C. _____

© 2016 Cengage Learning®. May not be scanned, copied or duplicated, or posted to a publicly accessible website, in whole or in part.

Name_____ Date_____

3.

A. _____
B. _____
C. _____

4.

A. _____
B. _____
C. _____

5.

A. _____
B. _____
C. _____

Name_____ Date_____

6.

A. _____
B. _____
C. _____

7.

A. _____
B. _____
C. _____

234 SECTION 7 Horses

Name_____ Date_____

EXERCISE 31-2 ANATOMY IDENTIFICATION EXERCISE

Knowledge of the basic exterior equine anatomy is important for anyone working with horses.

Directions: Label the following figure of a horse. Identify the part of anatomy at each line.

Name_____ Date_____

EXERCISE 31-3 UNSOUNDNESS ACTIVITY

The following table chart several unsound characteristics of horses. These blemishes can be aesthetic or impede performance of the horse and can even keep them from being registered.

Table for Exercise 31-3: Unsoundness Activity		
Unsoundness/Stable Vice	**Area Affected**	**Symptoms and Causes**
Heaving	Lungs	- Loss of elasticity in the lungs - Contraction of the flank muscles during expiration - Can be caused by a dusty atmosphere or hay
Bowed Tendons	Limbs	- Thickening of the back surface of the leg immediately above the fetlock
Bog Spavin	Limbs	- Soft fluctuating enlargement located at the upper part of the hock due to a distention of the joint capsule
Bucked Knee	Limbs	- Knee is in front of the plumb line
Camped Out	Limbs	- Knee is behind the plumb line
Contracted Heel	Limbs	- Back of the foot becomes narrower than normal
Founder/Laminitis	Limbs	- Inflammation of the sensitive laminae that attach the hoof to the fleshy portion of the foot - Happens when horses consume unlimited grain
Navicular Disease	Limbs	- Inflammation of the navicular bone and bursa
Wind Sucking		- Pressing the upper front teeth on an object, while pulling backward and sucking air into the stomach
Cribbing		- Horses grasp an object between their teeth and apply pressure, gradually gnawing away at the object
Weaving		- A rhythmical shifting of the weight from one front foot to the other. - Caused by an idleness in confined quarters
Roaring	Lungs	- Paralysis of the vocal cord nerve resulting in roaring or whistling while inhaling.
Blindness	Head	- Impaired vision, cloudiness of the cornea and white coloring.
Bad Mouth	Head	- Describes various jaw or tooth misalignments.
Poll Evil	Head	- A lesion or sore on the poll of the horse
Quidding	Head	- The horse drops food while eating due to mouth or teeth problems.
Fistulous Wither	Body	- Inflammation affecting the withers (similar to poll evil)
Sweeney	Body	- Muscle atrophy of the shoulder
Knocked Down Hip	Body	- A fracture of the external angle of the hip bone.

Name_____ Date_____

Activity: Using the chart, create an entertaining game that teaches these blemishes to the people playing the game. This game will be played by the class, and each student will facilitate the game he or she creates.

Game Suggestions (alternative games may be created if approved by the instructor)

Pictionary: Create pictures for 10 of the unsound characteristics. For each picture, there should be an information page behind it that shows the correct name of the blemish and the description. (This could be done by hand on flash cards or using a computer program and projection.)

Jeopardy: Create a jeopardy game using the unsound characteristics. There need to be at least four categories with five questions in each category. This game can be created on the computer. There are several jeopardy PowerPoint templates that can be located on the Internet. A handmade version of the game is also acceptable.

Wheel of Unsoundness: Create a spinning wheel with at least 15 unsound characteristics. The players will spin the wheel and answer with the description of whichever characteristic they land on. When making the game, make a page with the description of each blemish from the wheel. This key will be used to tell players if they are right or wrong. This can be made by hand or by using a template from the Internet.

Name_____ Date_____

CHAPTER 31 MATCHING ACTIVITY

Term

____ 1. light horses
____ 2. hand
____ 3. tobiano
____ 4. overo
____ 5. donkey
____ 6. jack
____ 7. jennet
____ 8. mule
____ 9. foal
____ 10. filly

Definition

a. a young horse of either sex up to 1 year of age

b. the male ass

c. the common name for the ass; the ass is smaller than the horse, has longer ears, and a short, erect mane

d. the offspring when a jack is crossed with a mare (female horse)

e. used mainly for riding, driving, and racing; measures 14–2 to 17 hands at the withers

f. a unit to measure horse height; each unit equals 4 in.

g. a type of paint horse whose head is marked in the same way as that of a solid-colored horse; the legs are white, at least below the knees and hocks; there are regular spots on the body

h. a female less than 3 years of age

i. female ass

j. paint horse with variable color head markings; the white usually does not cross the back between the withers and the tail; one or more legs are dark colored and the body markings are irregular and scattered

Name_____ Date_____

CHAPTER 31 LAB QUESTIONS

1. Describe the current uses of horses in the United States.

_____
_____
_____
_____
_____

2. What is the economic impact of the horse industry in the United States?

_____
_____
_____
_____
_____

3. What are the five basic colors of horses?

_____
_____
_____
_____
_____

Chapter 32

Feeding, Management, Housing, and Tack

INTRODUCTION

Horses are fed diets with a large percentage of roughage. Horses have a simple stomach rather than a rumen, so a horse owner must understand the digestive characteristics needed in horse feed. The amount of roughage and concentrate that a horse needs varies tremendously depending on the size, stage of growth, condition, and amount of work the horse needs to perform. Mature, idle horses that do not use a lot of energy can be fed on a ration composed solely of roughage. Grass and legume pastures can provide much of the roughage needed by a horse. For horses that are working or pregnant, or for growing foals, some types of concentrates are needed in the ration to provide adequate energy and protein.

Grains such as corn, milo, barley, and wheat can be used in a horse ration, but oats are the preferred grain. Soybean meal can be used as a supplement for protein, but supplements are not often needed. Many small horse operations purchase commercial protein supplements or complete pelleted rations. Three major concentrates that need to be part of a horse ration are salt, calcium, and phosphorus. Regular feeding and watering are important to managing the health of horses.

A proper breeding program is essential in many equine operations. Most mares are bred in the spring because they are more likely to conceive a foal. Pregnant mares need a proper feeding ration, exercise, and extra attention as they approach foaling. Mares should be monitored during foaling in case there is a need for assistance. Proper feeding and training of the foal is essential. Weaning at 4 to 6 months is a common practice.

All equine operations should be equipped with proper equipment for grooming, trimming hooves, saddling, and reigning horses. Tack is expensive and should be kept clean and in good repair. The Western and English saddles are the two most common types in use. Bridles and bits are selected on the basis of the use of the horse. This chapter requires the student to identify proper feed rations for horses at different stages of development as well as a hands-on horse tack lab.

240 SECTION 7 Horses

Name_____ Date_____

EXERCISE 32-1 FEEDING FOR APPROPRIATE SIZE AND AGE

When feeding horses, having a separate ration for animals at different stages of development is important. The energy and protein requirements for each stage vary, and it is important that a horse is not feed too much or too little of any nutrient. Use the Horse Feeding Guide and the Rations for Light Horses table from the textbook as a reference for this exercise.

Directions: For each of the following horse descriptions, suggest an appropriate ration that could be fed to each horse and how much to feed. The ingredients of the ration should be listed as a weight (lb) to be fed (daily allowance).

1. **A weanling colt weighing 600 lb**

    Ingredients: _____
    _____
    _____

    Daily allowance: _____
    _____
    _____

2. **A yearling weighing 900 lb**

    Ingredients: _____
    _____
    _____

    Daily allowance: _____
    _____
    _____

3. **A 1000-lb working horse being fed a conditioning ration**

    Ingredients: _____
    _____
    _____

    Daily allowance: _____
    _____
    _____

4. **A mare that is 8 months pregnant**

    Ingredients: _____
    _____
    _____

Name_____ Date_____

Daily allowance: _____
_____
_____

5. **A 1000-lb 2-year-old mare (not bred)**

   Ingredients: _____
   _____
   _____

   Daily allowance: _____
   _____
   _____

6. **A 1000-lb lactating mare**

   Ingredients: _____
   _____
   _____

   Daily allowance: _____
   _____
   _____

7. **A 1000-lb 22-year-old idle horse**

   Ingredients: _____
   _____
   _____

   Daily allowance: _____
   _____
   _____

8. **A 300-lb suckling foal**

   Ingredients: _____
   _____
   _____

   Daily allowance: _____
   _____
   _____

242 SECTION 7 Horses

Name_____ Date_____

EXERCISE 32-2 HORSE TACK IDENTIFICATION

Horse equipment (tack) is very expensive. The ability to identify and maintain horse tack is just as important as the ability to properly place and adjust the equipment on a horse. Part 1 is an identification activity, and Part 2 will require hands-on work tacking a horse.

Part 1

Directions: Label each arrow on the Figures 32-1 and 32-2. Tear the diagrams out and turn them into the instructor for grading.

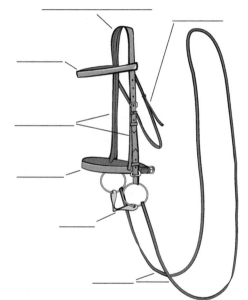

Figure 32-1   Diagram of a bridle.

Figure 32-2   English saddle diagrams.

© 2016 Cengage Learning®. May not be scanned, copied or duplicated, or posted to a publicly accessible website, in whole or in part.

Name_____ Date_____

Part 2

Directions: Perform the following tacking tasks.

Materials Needed: Demonstration horses, several styles of tack equipment (grooming equipment, saddles, girths, bridals, leg gear, and leather straps for tying demonstrations).

Task 1

Tying Skills: Using sample leather straps, practice each of the knots from Figure 32-3. Show each knot to the instructor for approval before moving on to the next style of knot.

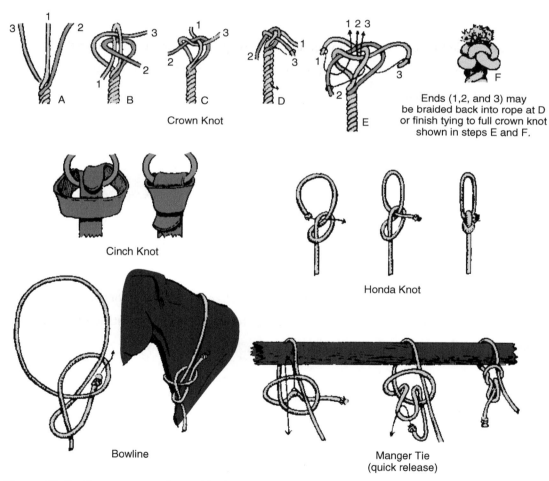

Figure 32-3 Knots commonly used by horse owners.

Name_____ Date_____

Task 2

Grooming: Tie the horse's lead to a rail and begin grooming practices. Curry comb, brush, and pick hooves with pick. Once the instructor indicates that the horse is properly groomed, move on to the next task.

Task 3

Saddling: Prepare the saddle to mount on the horse; lift one stirrup, and place the saddle on the horse in the proper location. Adjust all girths and stirrups to properly fit the horse and rider. Once the instructor indicates that the horse is properly saddled, move on to the next task.

Task 4

Bridle Adjustment: Choose the proper bridle and bit for the horse. Place the bit in the horse's mouth and the bridle over the muzzle and ears. Adjust the bridle so that it is not too tight or too loose. Once the instructor has approved the bridal, the horse is ready to ride.

Name_____ Date_____

CHAPTER 32 MATCHING ACTIVITY

Term **Definition**

____ 1. meconium a. used for tying or leading horses

____ 2. imprinting b. a person who works on horses' feet

____ 3. roached c. the feces that are impacted in the bowels during prenatal growth

____ 4. float d. the gear used to attach a horse or other draft animal to a load

____ 5. farrier e. its purpose is to hold the bit in the horse's mouth

____ 6. bridle f. a long-handled rasp with a guard to prevent injury to the horse during treatment

____ 7. bridle

____ 8. bit g. piece of hard material put in the horse's mouth to help control the horse

____ 9. halters h. a behavior-shaping process that takes place within the first 24 hours of a foal's life

____ 10. harness

i. the equipment used for riding and showing horses

j. is trimmed down to the crest of the neck, leaving the foretop and a wisp of mane at the withers

Name_____ Date_____

CHAPTER 32 LAB QUESTIONS

1. Explain how a horse's diet is different from other livestock discussed in this lab book.

   _____
   _____
   _____
   _____
   _____

2. How do the seasons affect horse reproduction?

   _____
   _____
   _____
   _____
   _____

3. Make a list of tack that should be kept when owning a horse for performance.

   _____
   _____
   _____
   _____
   _____

Chapter 33

Diseases and Parasites of Horses

INTRODUCTION

Parasites and equine disorders can have costly impacts on horses and horse owners. The symptoms of these health issues can cause loss of weight and loss in performance and can result in costly treatment methods and, in the worst case, death of the animal. Proper feeding and good management help reduce losses from diseases and parasites. Cleanliness and sanitation of the equine facilities and pastures are the most important aspects of a disease and parasite prevention program. Proper exercise and grooming are also necessary to help keep horses in good health and free from parasites and bacteria.

There are vaccines available for many types of equine diseases, but there are some that cannot be vaccinated against. The most serious diseases of horses are distemper, encephalomyelitis, equine infectious anemia, and equine influenza. When developing a vaccination program, a veterinarian familiar with the horse should be consulted. Knowledge of the diseases and symptoms is necessary for early detection. The longer the disease goes untreated, the lower the chance of successful treatment.

External and internal parasites pose a risk to horses every year. There are many different parasites that use horses as a host. The common external parasites of horses are flies, lice, mites, ringworms, and ticks. These parasites can cause digestion problems, irritation and swelling of skin, and constant discomfort to the horse. To control external parasites, there are several brands of insecticides that can be applied to eliminate the parasite.

More than 75 different species of internal parasites affect horses. The most serious internal parasites are strongyles, ascarids, pinworms, and bots. Heavy infestations of internal parasites lead to poor physical condition and even death of the horse. While horses of all ages are affected, young horses are more susceptible and seriously affected. For internal parasites, the most effective control is taking prevention measures based on the life cycle of the parasite.

The following exercises will develop knowledge of the various equine disorders and the treatment/prevention strategies available and provide a parasite control lab and worksheets.

248 SECTION 7 Horses

Name_____ Date_____

EXERCISE 33-1 EQUINE DISORDERS

Directions: For each of the following equine disorders, describe (1) the symptoms of the disorder, (2) prevention of the disorder, and (3) the treatment when symptoms are presented.

Anhydrosis
1. _____
2. _____
3. _____

Anthrax
1. _____
2. _____
3. _____

Azoturia (Monday morning sickness)
1. _____
2. _____
3. _____

Colic
1. _____
2. _____
3. _____

Cushing's syndrome
1. _____
2. _____
3. _____

Distemper (strangles)
1. _____
2. _____
3. _____

Name_____ Date_____

EXERCISE 33-2 PARASITE LIFE CYCLE LAB

Figure 33-1 presents an example of a horse parasite's life cycle. Understanding the life cycle is important in order to treat and prevent the infection of many types of parasites. In this exercise, complete the parasite identification activity, as well as the treatment and prevention charts.

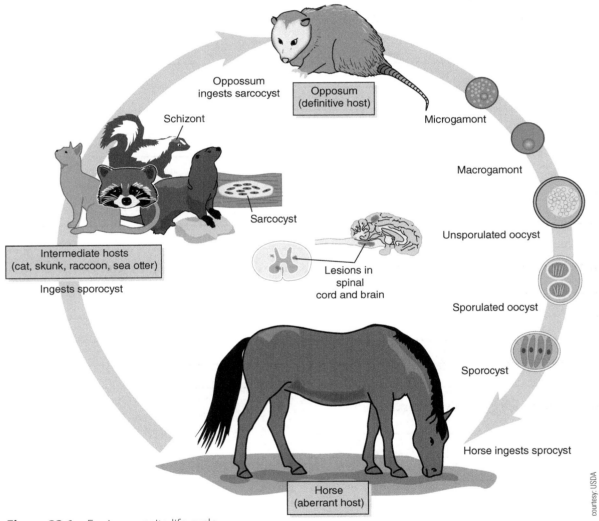

Figure 33-1 Equine parasite life cycle.

Directions: The instructor will divide the class into groups of two to three students each and then hand out index cards with a common parasite listed on each one. Examples are ascarids, small and large strongyles, and pinworms. Each group will need to answer the worksheet 1 questions based on the index cards. Next, the group should complete the parasite control worksheet 2.

© 2016 Cengage Learning®. May not be scanned, copied or duplicated, or posted to a publicly accessible website, in whole or in part.

Name_____ Date_____

Parasite Worksheet 1

The instructor will provide Playdoh. Present the findings to the class.

1. What is your parasite?
   _____
   _____
   _____

2. Describe your parasite. How big is it? What does it look like as an adult? Using Playdoh, make a model of your parasite to scale.
   _____
   _____
   _____
   _____
   _____

3. Where does your parasite live as a larvae? As an adult?
   _____
   _____
   _____

4. What organs and other internal structures does your parasite damage during its life cycle? How?
   _____
   _____
   _____
   _____
   _____

5. List or draw the life cycle of your parasite.
   _____
   _____
   _____
   _____

6. List five things you can do to keep help prevent your pony from getting parasites.
   _____
   _____
   _____
   _____

Name_____ Date_____

7. Why is it important to deworm a new horse in the barn? What method would you use to deworm a new horse?
_____
_____
_____
_____

Parasite Worksheet 2

Complete the first chart with examples of three different types of parasite control. In the second chart, schedule the use of each parasite control for five different groups of horses.

Class	Brand	Active Ingredient
Avermectins		
Benzimidazoles		
Tetrahydropyrimidines		

© 2016 Cengage Learning®. May not be scanned, copied or duplicated, or posted to a publicly accessible website, in whole or in part.

SECTION 7 Horses

Name_____ Date_____

Situation	Jan	Feb	Mar	Apr	May	June	Jul	Aug	Sep	Oct	Nov	Dec
A 10 horses/1000 acres												
B 20 horses/small pens												
C 2 horses/4 acres												
D 12 young horses/paddocks												
E 15 seniors												

A–Avermectin **B**–Benzimidazole **T**–Tetrahydropyrimidine

Name_____ Date_____

CHAPTER 33 MATCHING ACTIVITY

Term

____ 1. anhydrosis
____ 2. anthrax
____ 3. azoturia
____ 4. colic
____ 5. Cushing's syndrome
____ 6. distemper
____ 7. founder
____ 8. bot
____ 9. strongyle
____ 10. ascarid

Definition

a. a nutritional disorder that develops when a horse is put to work following a period of idleness

b. a type of parasitic roundworm that infests the intestines of animals

c. a disease caused by a small benign tumor in the pituitary gland

d. symptoms include high fever, loss of appetite, and depression; there is a pus-like discharge from the nose

e. flies that produce larvae that are parasites of horses

f. a condition in which horses do not sweat normally

g. a type of parasitic, blood-sucking worm that attacks the organs and tissues of animals

h. is not a specific disease, but rather a disease complex encompassing a wide range of conditions that affect the horse's digestive tract

i. symptoms include swelling of the sensitive laminae on one or more feet, lameness, fever, and sweating

j. a serious bacterial disease that is zoonotic

Name_____ Date_____

CHAPTER 33 LAB QUESTIONS

1. What are the results if a horse is untreated for parasites and disorders?
   _____
   _____
   _____
   _____

2. What is the most effective strategy to control disorders and parasites?
   _____
   _____
   _____
   _____

3. List the most common equine disorders.
   _____
   _____
   _____
   _____

4. How many species of parasites affect the horse?
   _____
   _____
   _____
   _____

Chapter 34

Training and Horsemanship

INTRODUCTION

Training a young horse is a delicate, slow, and detailed process. Each skill that riders usually take for granted must be trained from the beginning. Experienced trainers receive a premium fee for their services because working and training horses is a difficult job and requires knowledge of equine psychology, patience, and repetition. A horse that cannot be trained loses value at a fast rate. Understanding the reasons for horse behavior is helpful when training and riding horses. Horses have to move their heads to bring objects into focus so they naturally shy away at sudden movements. Horses have good memories, making it possible to train them to respond to commands, but negative experiences also remain in their memory and are hard to correct. Sensitive areas on the body of the horse are used to help control the animal, so a relationship of trust must be established between the trainer and horse.

Groundwork is the first and most essential part of a horse's training. Horses need to develop basic habits such as lifting their feet for farrier work and being led from place to place. Some basic groundwork that must be done when "breaking" or training a young horse includes haltering, leading, working with feet, saddling, longeing, long reining, neck reining, and pole work. Once the groundwork is established, training with the saddle may begin. The use of a hackamore and a bridle with a snaffle bit are strategies to maintain control of the untrained horse without causing pain or discomfort. When mounting a horse for the first time, special precaution must be taken to ensure safety to the horse and rider.

As the use of horses for work and transportation has decreased in U.S. history, horse equitation has gained in popularity. Equitation classes are judged on the performance of the rider. The rider must have skill in riding and controlling the horse. Position in the saddle, use of the hands, proper tack and dress, and the performance of the various gaits are the main points judged in equitation classes. In this exercise, complete the young horse training activity and explore various equitation events available for horse and rider to compete against other horse and rider pairs.

Name_____ Date_____

EXERCISE 34-1 TRAINING A YOUNG HORSE ACTIVITY

Young horses should be handled every day to repeat basic groundwork tasks. At the beginning, the time spent with young horses should not exceed 5 to 10 minutes for each training session.

Directions: For each of the training situations, describe three important training procedures required for proper training of the desired task. For each step, include a picture or drawing of the horse and trainer performing the task. In the space provided, draw each step or find a picture that can be placed in the space provided. Training procedures should include drawings or pictures.

Haltering
1. _____
2. _____
3. _____

Leading
1. _____
2. _____
3. _____

Working with the feet
1. _____
2. _____
3. _____

Longeing
1. _____
2. _____
3. _____

Saddling
1. _____
2. _____
3. _____

Name_____ Date_____

Using a hackamore and bridle
1. _____
2. _____
3. _____

Driving
1. _____
2. _____
3. _____

Name_____ Date_____

EXERCISE 34-2 EQUINE PERFORMANCE EVENTS

Equine performance events have grown in popularity as many horse owners compete as a hobby or pastime. These events serve as spectator events and a form of social interaction for riders and families that participate.

Directions: For each of the following equine events, (1) describe the objectives and rules of the event; (2) describe the appropriate equipment for the event, including tack and proper clothing for the rider; and (3) provide two examples of horse breeds that are commonly used for this event.

Horse evaluation

1. _____
2. _____
3. _____

English equitation

1. _____
2. _____
3. _____

Gymkhana

1. _____
2. _____
3. _____

Rodeos

1. _____
2. _____
3. _____

Name_____ Date_____

CHAPTER 34 MATCHING ACTIVITY

Term	**Definition**
____ 1. longeing	**a.** when the rider rises and sits in the saddle
____ 2. horsemanship	**b.** controlling the response of the horse by the weight of the rein against the neck
____ 3. neck-reining	**c.** training the horse at the end of a 25- to 30-ft line
____ 4. posting	**d.** the term used for games on horseback
____ 5. gymkhana	**e.** the art of riding a horse

Name_____ Date_____

CHAPTER 34   LAB QUESTIONS

1. What are the basic, natural behaviors of a horse that a trainer must understand?

   _____
   _____
   _____
   _____

2. What are some groundwork practices? Why are these important?

   _____
   _____
   _____
   _____

3. Describe the process of mounting a horse for the first time.

   _____
   _____
   _____
   _____

4. What is equitation? Why is it popular in the United States?

   _____
   _____
   _____
   _____

SECTION 8

Poultry

Chapter 35   Selection of Poultry   264

Chapter 36   Feeding, Management, Housing, and Equipment   270

Chapter 37   Diseases and Parasites of Poultry   277

Chapter 38   Marketing Poultry and Eggs   283

Chapter 35

Selection of Poultry

INTRODUCTION

Besides niche markets and small-scale operations, most of the poultry industry consists of large commercial production flocks. The three major types of poultry enterprises are egg production, meat production, and raising pullets for replacement. The majority of the poultry industry consists of chicken production—either broilers for meat or layers for egg production. Turkeys, ducks, and geese are commonly raised for meat in the United States and only make up a small part of the industry. A large percentage of the industry is vertically integrated in order for the production companies to control the product quality from production to the consumer.

The production of poultry has fluctuated over the past four or five decades but has seen a steady increase in recent years. Although the production of eggs has increased, the consumption per capita has decreased. Iowa and Ohio are the leading states for egg production. Georgia and Arkansas are the leading states in the production of chicken broilers. There has been a major increase in broiler production and per capita consumption of poultry since 1970.

Most chickens used for egg production in the United States are Leghorn strain crosses. Crossbred chickens are generally used for broiler production because of accelerated growth rates. In either section of the industry, the chickens are bred for the type of production desired. Niche market poultry producers utilize purebred chickens more often to differentiate their product. The breed most used for production turkeys is Broad Breasted Large Whites. The White Pekin duck is most commonly used for meat production, and the most popular breeds of geese are the Toulouse and the Embden, used primarily for meat production.

Poultry enterprises must purchase replacement birds on a regular basis as well a cull older birds that decrease in production. Most replacement birds are purchased from hatchery operations as day-old chicks, started birds, or ready-to-lay chickens. Culling is based on appearance and condition of the body. Pigmentation and molt are indicators of birds losing production quality for egg-laying production. This chapter explores the poultry industry as well as the anatomy of a chicken.

Name_____ Date_____

EXERCISE 35-1 GROWTH OF THE POULTRY INDUSTRY

The poultry industry has fluctuated over the past five decades as a result of many different factors. Over recent years, the trend in poultry has been an increased number of production chickens. In this exercise, assume the role of a poultry industry consulting firm. A client is looking to start up a poultry operation and needs assistance understanding the historical trends of the market, identifying the current market conditions, and the choosing the best location and type of enterprise to enter the industry. Groups of students will analyze the poultry market and suggest the best strategy for this client to enter the industry.

Directions: The instructor will divide the class into consulting teams. As a team, create a market analysis for the perspective poultry producer. The market analysis should include the following information:
- Describe the different poultry enterprises that are commonly used.
- Explain vertical integration.
- Describe the trends in poultry production and consumer consumption.
- Explain the recent trends for broilers and for egg production.
- Give a short description of the turkey industry trends for comparison.
- Suggest a location for the operation based on the records of production in various states.
- Describe the advantages and disadvantages to raising poultry.
- Outline the various breeds that can be used, and list their characteristics.
- Suggest sources where the client may obtain chickens for production.
- Summarize a suggested strategy for this client to enter the poultry industry.

The market analysis should be written as a proposal to be handed into the instructor or as a presented proposal using a computer-based presentation program.

EXERCISE 35-2 CHICKEN ANATOMY ACTIVITY

Locating and learning the poultry parts on a live bird is an important skill in the industry. Before real birds are used, this exercise will use a diagram to identify the parts of a chicken in Part 1, and Part 2 includes a chicken wing dissection to better understand the internal anatomy.

Part 1

Directions: Using the word bank, identify the external parts of a chicken.

Word Bank:

Beak	Breast	Comb	Ear Lobe
Hock	Eye	Shank	Shoulder
Toes	Vent	Waddle	

Name_____ Date_____

Part 2: Chicken Wing Dissection

Through this dissection activity, make sure to observe the muscle, bones, and blood vessels that make up a chicken's wing. The instructor will provide all of the needed materials for the dissection.

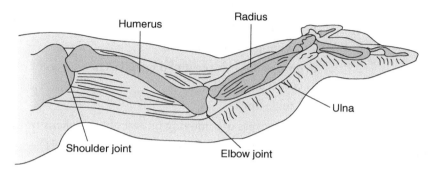

Dissection instructions

1. Rinse the chicken wing under cool water and then dry it off
2. Using scissors, cut the skin down the middle of the wing from where it was removed from the body toward the tip of the wing. (The scissors should cut the skin but not the muscle.)
3. Next, cut a line across the first line, approximately in the middle of the first cut.
4. Peel back the layers of skin to reveal the muscle tendons and bones.
5. Observe all of the internal parts of the chicken wing, and complete the following table with the information that is gathered.

Tissue	Description (color, texture, etc.)	Tissues it is attached to
Skin		
Fat		
Muscle		
Tendon		
Ligament		
Cartilage		

Conclusion

Based on the observations, explain the roles of muscles, bones, tendons, and joints in the movement of the lower chicken wing.

CHAPTER 35 MATCHING ACTIVITY

Term

____ 1. vertical integration
____ 2. culling
____ 3. variety
____ 4. cross-mating
____ 5. hybrid
____ 6. comb
____ 7. pinfeather
____ 8. drake
____ 9. straight-run chicks
____ 10. pullet

Definition

a. breeds that are a direct result of breed crossing; crosses are also popular for egg production

b. crossing two or more strains within the same breed

c. a young female chicken

d. groups of birds separated by differences in color of plumage, type of comb, and size

e. a male duck

f. the process of removing undesirable chickens from the flock

g. about one-half pullets and one-half cockerels

h. where two or more steps of production, marketing, and processing are linked together

i. a feather that is not fully developed

j. examples include single, rose, pea, cushion, buttercup, strawberry, and V-shaped, and walnut

Name_____ Date_____

CHAPTER 35 LAB QUESTIONS

1. What recent changes have occurred in the poultry industry?

   _____
   _____
   _____
   _____

2. Explain the differences between a broiler and a laying hen.

   _____
   _____
   _____
   _____

3. What are the leading states of the United States in the production of broilers? Layers?

   _____
   _____
   _____
   _____

Chapter 36

Feeding, Management, Housing, and Equipment

INTRODUCTION

Poultry require carbohydrates, protein, minerals, vitamins, and water in order to grow and remain healthy. The highest cost in poultry production is the expense of feed. Smaller flock owners generally find it easier to use a complete commercial feed, while large commercial flock owners usually use some system of mixing ingredients to make a complete feed. Two ingredients commonly added to chicken feed are grit and oyster shells. Grit is used to help the bird grind the feed in the gizzard, and oyster shell provides calcium for egg production.

Water is an essential part of a poultry operation. Controlling the amount and quality of water can affect the profitability of the flock directly. Understanding the expected water consumption rates and the acceptable levels of compounds in the water is essential when raising poultry. Light and temperature have a significant effect on the growth of poultry. Light affects the sexual maturity and egg production rates of poultry. Proper temperatures are required for successful brooding, meat production and egg production. Many poultry production buildings are equipped with tunnel ventilation fans to maintain an adequate temperature.

Poultry may be brooded on the floor or in cages. Chickens raised for meat production are usually raised in confinement, unless supplying a niche market such a free range chicken. Laying flocks may be handled using floor systems or in cages. Automation has been utilized in a majority of laying operations to reduce labor and human error. Eggs must be carefully handled to maintain quality. Chickens raised for meat production are usually raised in confinement.

Housing and equipment requirements differ greatly depending on the size, age, species, breed, and purpose that poultry are raised for. Turkeys, geese, and ducks generally require less investment in facilities and equipment. Automatic or semiautomatic feeding, watering, and cleaning equipment are frequently used in the commercial poultry industry. The standards demanded through vertical integration require most poultry producers to eliminate any areas of inefficiency in the operation.

This chapter provides two exercises that develop an understanding of poultry requirements.

Name_____ Date_____

EXERCISE 36-1 WATER CONSUMPTION LAB

Water is a critical nutrient that receives little attention until a problem arises. Not only should producers make an effort to provide water in adequate quantity, but they should also know what is in the water that will be consumed by the poultry. In this three-part exercise, students will study the water consumption levels of birds at different sizes and stages of life, as well as analyze the quality of the water being provided to the poultry.

Part 1

Directions: Use information from the textbook to determine how much water should be consumed each day by different groups of chickens. For each of the following types of chicken and groups, determine the needed water consumption to keep the entire flock hydrated.

A flock of 500, 1-week-old broilers:
_____
_____

A flock of 100, 4-week-old white leg horn hens:
_____
_____

A flock of 400, 1-week-old brown hens:
_____
_____

A flock of 5000, 8-week-old broilers:
_____
_____

A flock of 2000, 16-week-old white leghorn hens:
_____
_____

A flock of 1500, 20-week-old brown hens:
_____
_____

A flock of 750, 4-week-old broilers:
_____
_____

A flock of 2500, 1-week-old brown hens:
_____
_____

Name_____ Date_____

Part 2

The instructor will provide several samples of live birds. There will be an assortment of birds of different ages, sizes and breeds. A total of six different birds should be used. Separate each bird into a distinct area with its own feed and water. Monitor the feed and water intake of each bird. Use a watering container with measurement markings for easy data recording.

Directions: Monitor and record the data for each bird and answer the following questions.

1. Determine the bird that requires the most water intake.
_____
_____
_____

2. Why does the bird with the highest intake consume so much water?
_____
_____
_____

3. How much feed did the bird consume?
_____
_____
_____

4. What is the bird's feed-to-water ratio?
_____
_____
_____

5. What were the environmental factors that could have affected the water consumption rates?
_____
_____
_____
_____

Name_____ Date_____

Part 3

The quality of water provided to poultry is just as important as providing the adequate amount. In this exercise, use a standard water quality test provided by the instructor to determine the level of various substances in the water.

Directions: Table 36-1 is a guideline for acceptable levels of compounds found in drinking water. Collect several samples of water from the waterers located on at a sample poultry operation. Compare the levels of compounds in the drinking water to the levels given in the guidelines. Create an analysis report and determine if the sample water is acceptable for poultry to consume. Present your findings to the instructor and the class.

Table 36-1

Characteristic or Mineral	Maximum Acceptable Levels
pH	6.0–8.0
Hardness	110 ppm
Naturally Occurring Compounds	
Calcium	500 ppm
Chloride	250 ppm
Copper	0.6 ppm
Iron	0.03 ppm
Magnesium	125 ppm
Manganese	0.05 ppm
Nitrate	25 ppm
Phosphorus	0.1 ppm
Potassium	500 ppm
Sodium	50 ppm

Name_____ Date_____

EXERCISE 36-2 MANAGEMENT LAB: BROILERS vs. LAYERS

Managing chicken operations for meat production and egg production requires two very different management strategies. The requirements for feed, housing, equipment, animal health, labor, growth, and product processing are completely different in laying operations compared to a meat production operation.

Directions: In the space provided, highlight the major differences in management requirements for an egg production operation and a meat production operation. Include specific requirements recommended in the textbook. Also include any similarities in requirements.

Egg production system requirements:

_____
_____
_____
_____
_____
_____
_____

Meat production system requirements:

_____
_____
_____
_____
_____
_____
_____

List any similarities between production systems:

_____
_____
_____
_____
_____
_____
_____

Name_____ Date_____

CHAPTER 36 MATCHING ACTIVITY

Term

____ 1. ad libitum
____ 2. grit
____ 3. phase feeding
____ 4. feeding efficiency
____ 5. poult
____ 6. hover guard
____ 7. trimming
____ 8. foot-candle
____ 9. capon
____ 10. broodiness

Definition

a. the number of pounds (kilograms) of feed required to produce a dozen eggs

b. a male chicken that has been surgically castrated

c. a young turkey

d. when chickens are given all they will eat

e. the cutting off of one-third to one-half of the upper beak and one-fourth of the lower beak

f. a unit of luminance on a surface that is 1 ft from a point source of a candle or light

g. used for the first week of brooding to prevent the chicks from wandering away from the heat and becoming chilled

h. a system of making specific feeds to be used to meet the changing nutritional requirements of chickens

i. when a hen stops laying eggs and wants to sit on a nest of eggs to hatch them

j. small particles of granite; comes in small, medium, and large sizes for use with chicks, growing chickens, and adult chickens, respectively

Name_____ Date_____

CHAPTER 36 LAB QUESTIONS

1. What are the major requirements for raising poultry?
   _____
   _____
   _____
   _____

2. How does light affect poultry production?
   _____
   _____
   _____
   _____

3. What are three types of confinement structures for poultry? Explain each one.
   _____
   _____
   _____
   _____

Chapter 37

Diseases and Parasites of Poultry

INTRODUCTION

In order to maintain a good health program for poultry, a producer must focus on prevention of disease through various methods. The last resort should be treatment, which can be costly and is never guaranteed to help the birds recover. The best way to control disease and the infestation of parasites is to create a prevention plan. Prevention involves sanitation, good management, vaccination, and control to stop and treat disease outbreaks. When poultry animals become sick, they rarely recover. Treatment is only effective when the sickness is caught and treated early. Without a proper treatment plan, the profits of a poultry operation decrease significantly. The involvement of a veterinarian is recommended when developing a treatment plan or diagnosing an observed disease or parasite.

To properly sanitize during disease prevention and control, all poultry buildings and equipment should be disinfected and scrubbed free of any lingering litter from previous groups of birds. Allowing a dormant period for the housing and equipment to dry out between poultry groups is also recommended. Any birds that become infected should be removed from the flock and isolated where they cannot come in contact with healthy birds. Any dead birds should be disposed of, and manure from infected birds should not come into contact with healthy stock.

Other methods to prevent the outbreak of disease include vaccinating young birds and checking daily for signs of disease or parasite. A drop in production or an increase in the death rate may indicate the presence of a disease. Even veterinarians cannot always accurately diagnose an illness in poultry, so they may rely on laboratory test to assist in a diagnosis.

There are many diseases that affect poultry. A large number have similar symptoms and effects, and there are several medicines that treat a wide variety of sickness. Some diseases are specific to the species of poultry, while many sicknesses are common to other poultry species. All poultry birds are affected by mites and lice if not properly treated. Maintaining a sanitary environment and the proper use of insecticides can help control the spread of these parasites.

Name_____ Date_____

EXERCISE 37-1: DISEASE AND PARASITE CONTROL PROGRAM ACTIVITY

Understanding medication labels and properly administering them to the animals is an important skill. In this exercise, students will calculate the first day that any chickens can safely be marketed for food, calculate the correct amount of medication to mix with water when making stock solution, and calculate the amount of stock solution to mix with the chickens' drinking water.

Directions: Read the label instructions provided in Figure 37-1. Use the label and the following situation to answer all the questions about administering the medication.

Carramycin-152
(oxytetracycline HCl as soluble water)
For control and treatment of specific disease in poultry, cattle, swine and sheep.

CAREFULLY READ ALL DIRECTIONS BEFORE USING THIS PRODUCT.
Soluble Powder for Use in Drinking Water Only.

Active Ingredients: Carramycin-152 is a broad spectrum **antibiotic**. This 4.78 oz packet contains 102.4 grams oxytetracycline HCl (after mixing with clean, fresh water-512 gallons containing 200 mg oxytetracyline HCl per gallon. 256 gallons containing 400 mg oxytetracycline HCl per gallon: 128 gallons containing 800 mg oxytetracycline HCl per gallon).

Indications: For control of **poultry** diseases caused by organisms susceptible to oxytetracycline.

Recommended Dosage
Add the following amount to two (2) gallons of fresh, clean water to make a stock solution.
Mix one (1) ounce solution per one (1) gallon drinking water.

		Dosage	Packs/2 Gallons Stock solution
Chickens	Infectious synovitis	200 mg/gal	1/2
	Chronic respiratory disease	800 mg/gal	2
	Fowl Cholera	800 mg/gal	2
Turkeys	Infectious synovitis	400 mg/gal	1
	Hexamitiasis	200 mg/gal	1/2

Cautions: 1. Carramycin-152 is for use in flock drinking water only. 2. Medicate continuously at the first clinical signs of disease and continue for 7 to 14 days. If improvement is not noticed within 24 to 48 hours, consult a veterinarian or diagnostic laboratory to determine diagnosis and advice on dosage. 3. Use as sole source of oxytetracycline. Do not use for more than 14 consecutive day in chickens and turkeys or five (5) consecutive days in cattle, sheep or swine. 4. Carramycin-152 is to be stored below 77°F. 5. The concentration of drug required in the medicated water must be adequate to compensate for variations in age of the animal, feed consumption rate, and the environmental temperature and humidity-each of which affects water consumption.

Warning: Do not administer to chickens, turkeys, wine, cattle, or sheep within five (5) days of slaughter. Do not administer to chickens or turkeys producing eggs for human consumption.

How Supplied: Carramycin-152 soluble powder is available in packets of 4.78 oz.

Livestock Drug – *not for human use*   Distributed by
KEEP OUT OF REACH OF CHILDREN   **Livestock Health, Inc.**

Source: Poultry Learning Laboratory Kit. © 1998, Curriculum Materials Service, Ohio State University

Figure 37-1 Example of a pharmaceutical label.

Name_____ Date_____

Situation

As the owner of a poultry meat production operation, you care for the health of 3,200 red broilers. The flock health manager reports that most of the chickens are coughing and having difficulty breathing. After calling the veterinarian in to look at the flock, they are diagnosed with chronic respiratory infection. The good news is that this is a treatable infection. The veterinarian sells you a water-soluble medication that can be administered to the flock through the drinking water. The medicine is Carramycin-152. You instruct the manager to immediately begin mixing the powder into the water and keep track of the time period the medicine is administered; the start date is May 28. The veterinarian recommended treating the birds with the soluble powder mixed in the drinking water for 10 days. The directions on the medication label direct you to mix 800 mg of Carramycin for every gallon of water. The flock of broilers drinks a total of 256 gallons of water each day; for each gallon of water, mix one ounce of the soluble powder.

Medication Questions

1. How many packs of this medicine should be mixed with 2 gallons of water?
   _____
   _____
   _____

2. How much soluble powder is needed for 10 gallons of water?
   _____
   _____
   _____

3. How many packs of Carramycin will be needed to treat the broilers for 10 days?
   _____
   _____
   _____

4. How many ounces of the soluble powder will be needed each day?
   _____
   _____
   _____

Name_____ Date_____

5. How many milligrams of the medicine will each chicken in the flock receive?

_____

_____

_____

6. What is the first day the broiler could be slaughtered when all of the residue from the oxytetracycline are gone? Give a date.

_____

_____

_____

Name_____ Date_____

CHAPTER 37 MATCHING ACTIVITY

Term

____ 1. intranasal vaccination
____ 2. intraocular vaccination
____ 3. wing web vaccination
____ 4. synovia
____ 5. bacterin

Definition

a. the process of injecting the vaccine into the skin on the underside of the wing web at the elbow
b. a transparent lubricating fluid in the joints of poultry
c. placement of a vaccine directly into the nose opening
d. a vaccine made with weakened bacteria
e. placement of the vaccine directly into the eye

Name_____ Date_____

CHAPTER 37 LAB QUESTIONS

1. What are some of the major causes of disease and parasites in poultry?
_____
_____
_____
_____

2. What is the most effective strategy to control infection of disease and parasites?
_____
_____
_____
_____

3. List the most common diseases in poultry.
_____
_____
_____
_____

4. What are the most common parasites to affect poultry?
_____
_____
_____
_____

5. Describe the life cycle of parasites and how they enter the body of the animal.
_____
_____
_____
_____

Chapter 38

Marketing Poultry and Eggs

INTRODUCTION

The poultry market has fluctuated somewhat over the past 40 years, but in general, the consumption of broilers and turkeys has grown in an upward trend. The consumption of eggs has been decreasing over this period, but the trend has been rebounding in recent years. Long-term trends in prices show variations resulting from supply and demand and the influence of vertical integration on the market. Vertical integration is a system in which the company that markets the poultry also owns the poultry through the entire process of raising from chick to market animal.

Most of the large chicken brands that are recognizable across the nation in supermarkets are using vertical integration. In the United States, 99 percent of the broilers are produced under some type of vertical integration contract. At the large volumes that these integrated contracts are marketing, the poultry must be carefully handled during every step to avoid injuries and bruises. Most of the turkeys in the country are marketed through integrated contracts or cooperatives, while ducks and geese are more commonly processed on the farm and directly marketed to consumers.

Each species of poultry has carcass classes and grades to assess the quality of meat. Classes are based on age, tenderness of meat, smoothness of skin, and hardness of breastbone cartilage. These "ready-to-cook" carcasses are classed and then graded. The three are known as A, B, and C. These grades and all standards for grading poultry are found in the U.S. Department of Agriculture (USDA) Poultry Grading Manual.

A large section of the poultry egg industry is marketed though vertical integration as well. The rest of the egg market includes sale to local buyers, produce dealers, cooperatives, or direct sales to consumers. Eggs are classed and graded to measure the quality. Eggs must be handled carefully to maintain high quality because defects due to handling can decrease the value of an egg or even render it unsellable. The USDA also sets the standards for weight classes and grades of eggs. The egg classes are based on the weight in ounces per dozen eggs. The grades are based on shell quality, size of the air cell, firmness of the egg white, and the presence of defects in the yolk. The ability to grade and class eggs is necessary in the poultry industry.

Candling eggs to determine their grade.

Name_____ Date_____

EXERCISE 38-1 EGG GRADING LAB

Directions: Answer all of the following questions about quality grading eggs. Once the questions are complete, the instructor will provide a sample of various qualities of eggs. Use Table 38.1 in the textbook for quality grade specifications. Grade each sample egg, and explain why it received that grade.

Grading cartons of eggs

1. Name three factors used in grading a carton of eggs.
   _____
   _____
   _____
   _____

2. What is the difference between a checked and cracked egg?
   _____
   _____
   _____
   _____

3. What is a body check?
   _____
   _____
   _____
   _____

4. Grade example cartons of eggs provided by the instructor, and record the grades and reasons.
   _____
   _____
   _____
   _____
   _____

Name _____ Date _____

Interior egg evaluation

1. What process is used to evaluate the interior of an egg?
   _____
   _____
   _____
   _____

2. What popular comparison is used when determining the size of egg's air sac?
   _____
   _____
   _____
   _____

3. What common defects are evaluated when grading the interior of an egg?
   _____
   _____
   _____
   _____

4. Describe the characteristics of the yolk from a freshly laid egg.
   _____
   _____
   _____
   _____

5. Candle several sample eggs and determine the interior quality of the egg.
   _____
   _____
   _____
   _____

6. Determine the grade of egg based on its break-out quality.
   _____
   _____
   _____
   _____

Name_____ Date_____

Shell evaluation

1. What are the four grades of egg shells?

   _____
   _____
   _____

2. Describe the six factors used to determine egg shell quality.

   _____
   _____
   _____
   _____

3. What will automatically cause an egg to receive a grade of B?

   _____
   _____
   _____
   _____

4. What two defects cause a grade of dirty?

   _____
   _____
   _____
   _____

5. Grade the shells of all sample eggs. Explain the reasons for each grade.

   _____
   _____
   _____
   _____

Name_____ Date_____

CHAPTER 38 MATCHING ACTIVITY

Term   **Definition**

____ 1. viscera   a. eggs that are examined by using a high-intensity light
____ 2. green geese   b. internal poultry organs
____ 3. candling   c. geese that are full fed for fast growth

Name_____ Date_____

CHAPTER 38 LAB QUESTIONS

1. Describe the long-term trends in the poultry market.

   _____
   _____
   _____
   _____
   _____

2. What system is used for a majority of poultry marketing?

   _____
   _____
   _____
   _____
   _____

3. What qualities are "ready-to-cook" poultry carcasses graded upon?

   _____
   _____
   _____
   _____
   _____

SECTION 9

Dairy Cattle

Chapter 39	Breeds of Dairy Cattle	292
Chapter 40	Selecting and Judging Dairy Cattle	299
Chapter 41	Feeding Dairy Cattle	304
Chapter 42	Management of the Dairy Herd	311
Chapter 43	Milking Management	317
Chapter 44	Dairy Herd Health	322
Chapter 45	Dairy Housing and Equipment	328
Chapter 46	Marketing Milk	333

Chapter 39

Breeds of Dairy Cattle

INTRODUCTION

Many of the dairy breeds used in modern agriculture had their origins in Europe. The cattle were selected for the most desirable traits for the area. The most desirable animals were kept as breeding stock. The development of dairy breeds has continued into the current dairy industry. Herds that have been managed for generations have achieved a high level of genetics that receive premium value. Dairy cows that produce large amounts of milk or high levels of milk components are desired. Cows are constantly producing milk during lactation, so they are milked two to three times per day, 7 days a week, continuously.

The dairy industry is very difficult—and labor intensive. It also requires a large capital investment to begin a dairy operation; many commercial operations are generational family farms or large farms owned by investors. Extensive training is preferred for those managing dairy cattle. Most dairy farms hire a herd health manager and a college education is desirable.

The number of dairy farms in the United States has been declining in recent years, but the dairy cattle numbers have recently been increasing slightly. The trend has been for small herds of dairy cattle to leave the industry and the large commercial dairy farms to expand and grow. The average size of the dairy farm has increased due to this trend. Improvements in herd health, genetics, and milking equipment technology have caused the milk produced per cow to increase as well.

The seven major breeds of dairy cattle in the United States are Ayrshire, Brown Swiss, Guernsey, Holstein-Friesian, Jersey, Milking Shorthorn, and Red and White. There are breed associations for the purpose of registering purebred dairy cattle. Registry ensures the pedigree of the animal as well as increases its genetic value. Dairy cattle have been developed over the generations by selecting for quantity and persistence of milk production. Each breed is selected based on the characteristics commonly found in the breed, but when interested in milk production, a farmer should milk any breed of cattle that is profitable. The following exercises are designed to learn basic breed identification skills and an understanding of the dairy industry.

Name_____ Date_____

EXERCISE 39-1   TRENDS IN DAIRY PRODUCTION ACTIVITY

Create a timeline that highlights the various trends in the dairy industry using the graphs in Chapter 39 of the textbook. The timeline should track the changes in the dairy industry that have taken place from 1975 to 2010. The timeline should describe the conditions of the dairy industry during that year. The categories to include on the timeline are (1) the milk cow numbers, (2) total milk production, (3) milk production per cow, (4) production of milk fat per cow, (5) the consumption of fluid milk and cream, (6) consumption of milk and nonfat dry milk, (7) consumption of cheese and ice cream, and (8) the consumption of dry whole milk.

Directions: Create a timeline of the changes in the dairy industry since 1975. Include each category of dairy industry measurements that apply. Include the known reasons for the changes that occur.

Name_____ Date_____

Timeline of Dairy Industry

Date	Changes in the Industry (industry data)

CHAPTER 39   Breeds of Dairy Cattle   295

Name_____   Date_____

EXERCISE 39-2   BREED COMPARISON LAB

Part 1:

Directions: Each of the following pictures show a different dairy cattle breed. For each picture, identify the breed and write a brief history and description of the breed. Include any unique characteristics of the breed.

1.

   a. Breed: _____
   b. History: _____
   c. Description: _____

2.

   a. Breed: _____
   b. History: _____
   c. Description: _____

3.

   a. Breed: _____
   b. History: _____
   c. Description: _____

© 2016 Cengage Learning®. May not be scanned, copied or duplicated, or posted to a publicly accessible website, in whole or in part.

Name_____ Date_____

4.

a. Breed: _____
b. History: _____
c. Description: _____

5.

a. Breed: _____
b. History: _____
c. Description: _____

Part 2:

Directions: Compare breeds based on their use in production. Determine which breeds would be suitable to (1) large commercial operations that milk three times per day, (2) niche farms that produce alternative dairy products, (3) registered purebred herds, or (4) small-scale farms that sell dairy products directly to the consumer,

Brown Swiss

Holstein-Friesian

Ayrshire

Guernsey

Red and White

Jersey

Name_____ Date_____

CHAPTER 39 MATCHING ACTIVITY

Term

____ 1. Grade A milk
____ 2. Grade B milk
____ 3. registered

Definition

a. animals that meet the requirements of the breed association and are recorded in the herd book of the association

b. milk that is produced on farms that have been certified to meet certain minimum standards

c. milk that is produced under conditions that are less controlled than those for producing Grade A milk

Name_____ Date_____

CHAPTER 39 LAB QUESTIONS

1. What qualities of milk are desired other than quantity?

   _____
   _____
   _____
   _____

2. What is the difference between Grade A and Grade B milk?

   _____
   _____
   _____
   _____

3. What are the desirable qualities of a milking shorthorn?

   _____
   _____
   _____
   _____

4. What would be considered a small dairy farm in the United States?

   _____
   _____
   _____
   _____

Chapter 40
Selecting and Judging Dairy Cattle

INTRODUCTION

Judging dairy cattle gives a general idea of the qualities that a cow has. The more judging events an animal wins or performs well in, the more valuable its genetics will be. A good judge learns the parts of the animal and develops a logical system for evaluating the animal using the Dairy Cow Unified Scorecard.

Physical appearance, health, milk production records, and pedigree are important factors to consider when selecting dairy animals for breeding and production. The mammary system is one of the most important parts of the dairy cow and carries the most weight when scoring and evaluating the cow. These traits help determine the milk production of a cow over a long period of time.

When dairy cows are being evaluated, a judge should have a routine for visual inspection to ensure an accurate comparison between cattle. A dairy cow should be observed from the side, rear, and front. A judge should have experience and credibility in the cattle industry in order to build a reputation among dairy producers. The judge should give a presentation of oral reasons and use the proper terms to describe cattle conformation.

Producers have many strategies to best prepare a cow for a dairy cattle show. Feeding and water strategies prior to the judging event can affect the way the cow looks and scores. Clipping a cow to highlight the positive qualities, such as a straight topline, is a strategy used by most producers.

The exercise in this chapter is designed to develop skill using the Dairy Cow Unified Scorecard to evaluate cattle.

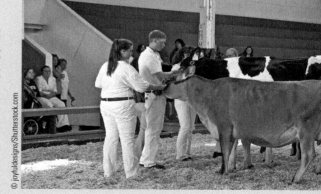

Youth cattle show.

300 SECTION 9 Dairy Cattle

Name_____ Date_____

EXERCISE 40-1 DAIRY COW SCORECARD

When evaluating dairy for quality of conformation and breed traits, the Dairy Cow Unified Scorecard is used. In Part 1 of this exercise, identify the general desired characteristics that should be observed. In Part 2, describe the desired characteristics of each breed based on the scorecard. Finally, in Part 3, evaluate a group of dairy cattle using the scorecard. Use the unified scorecard located in the textbook for reference.

Part 1

Directions: Identify the desired characteristics for each of the areas of evaluation.

1. General appearance:

_____
_____

2. Dairy character:

_____
_____

3. Body character:

_____
_____

4. Udder:

_____
_____

Part 2

Directions: Identify the desired characteristics of each of the following breeds and the factors to be evaluated.

Ayrshire:

_____
_____

Holstein:

_____
_____

Name_____ Date_____

Milking Shorthorn:
_____
_____

Brown Swiss:
_____
_____

Guernsey:
_____
_____

Part 3

Directions: The instructor will provide a situation to evaluate a group of lactating dairy cows. Score each cow based upon the Dairy Cow Unified Scorecard, give an explanation for each score, and place the cattle in order based on the score they receive.

Record evaluation results here:

Name_____ Date_____

CHAPTER 40 — MATCHING ACTIVITY

Term   **Definition**

____ 1. pedigree   **a.** an udder that is still firm after milking

____ 2. linear classification   **b.** the record of the animal's ancestors

____ 3. meaty   **c.** a modification of type classification that utilizes a computer program to score dairy cattle on a number of individual traits

Name _____ Date _____

CHAPTER 40 LAB QUESTIONS

1. Describe the term *dairy* and how it is judged on a dairy cow.

   _____
   _____
   _____
   _____

2. Describe the steps that should be taken when judging dairy cattle.

   _____
   _____
   _____
   _____

3. Make a list of 20 terms that may be used when describing dairy cattle with oral reasons.

   _____
   _____
   _____
   _____

Chapter 41
Feeding Dairy Cattle

INTRODUCTION

Approximately half of the cost of producing milk is the cost of feed. A total mixed ration allows a dairy cow to reach her full potential for milk production and genetic potential. A good, high-quality roughage is the basis of a ration, and grain and protein supplements make up the other portion of the animal feed. Many dairy farmers group cattle according to similar sizes and stages of production. This allows the farmer to mix a ration for each group that targets the needed energy requirements. This is much more efficient than trying to meet individual cows' nutritional requirements.

Hay, pasture, green chop, silage, and haylage are commonly used for roughage in dairy rations. Alfalfa hay and corn silage are two of the most common roughages used because they are the most economically produced. Corn and oats are the major grains used in the dairy industry. Protein supplements are essential in a total mixed ration (TMR); some common examples of protein supplements are soybean oil meal, linseed meal, and cottonseed.

Dairy producers may use other products in the TMR. The use of these products depends on the availability of the product, the experience the farmer has had in the past, the price of the product, and the affect it has on milk production. A variety of by-product and other processed feeds may be used, including alfalfa meal, beet pulp, brewer's grain, and urea as a source of nitrogen.

The inclusion of minerals in the ration is very important for milk production and milk quality. Calcium, phosphorus, salt, and minerals are important and needed when balancing dairy rations. A vitamin commercial mix is usually mixed into the TMR. Vitamins A, D, and E are important components of a dairy ration. The most limiting factor in milk production is usually a shortage of energy in the ration. The stages of lactation, the size of the cow, and the amount of milk being produced per day are all factors used to determine the nutritional requirements.

A dairy calf is taken from the cow soon after birth. Its navel is dipped in iodine, and it is fed colostrum from the mother. Colostrum is the first milk produced by the fresh cow. It contains antibodies that help protect the calf from disease.

Complete the following exercises on dairy feeding strategies.

Name_____ Date_____

EXERCISE 41-1 PEARSON'S SQUARE

The use of the Pearson's square was explained in detail in Chapter 8. Use the Pearson's square method for mixing ingredients in a TMR.

Directions: For each of the following feed scenarios, determine the protein values needed for a productive TMR. Show all work used to calculate the ration.

1. Create 1 ton of a 16 percent protein ration using corn (10.9 percent) and soybean meal (41 percent). How much of each will have to be mixed together?

_____

_____

_____

2. Make 1 ton of a 16 percent protein ration using alfalfa hay (18.7 percent) and oats (13.3 percent). How much of each will have to be mixed together?

_____

_____

_____

3. Increase the protein levels and make 1 ton of a 23 percent ration using oats (13.3 percent) and cottonseed meal (41 percent). How much of each will have to be used?

_____

_____

_____

4. Make a ration that is 22 percent protein. Use corn (10.9 percent) and dehydrated skimmed milk (35.8 percent). How much of each will have to be used to make 1 ton of this ration?

_____

_____

_____

5. Make 1 ton of a 20 percent protein ration. Use sorghum (12.4 percent) and soybean oil meal (41 percent). How much of each will have to be used in the ration?

_____

_____

_____

6. Mix a ration for dry cows that is 10 percent protein using prairie hay (5.8 percent) and a commercial supplement (36 percent) How much prairie hay and how much supplement will have to be used to make 1 ton of this feed?

_____

_____

_____

Name_____ Date_____

EXERCISE 41-2 TOTAL MIXED RATION LAB

Part 1 in this exercise will review the ability to read total mixed ration values, and Part 2 will practice the use of the fist test to determine the moisture of a TMR.

Part 1

Directions: For each description of a cow and the major ingredients available, list the ingredients and values for a total mixed ration. (Use Table 41-7 from the textbook.)

Cow 1

Description: Holstein heifer; 16 months; 900 lb; alfalfa mid-bloom

_____
_____
_____
_____
_____
_____
_____
_____

Cow 2

Description: Guernsey; 21 months; 1100 lb; grass

_____
_____
_____
_____
_____
_____
_____
_____

Name_____ Date_____

Cow 3

Description: Jersey; 5 month; 300 lb; budding alfalfa

_____
_____
_____
_____
_____
_____
_____
_____

Part 2

Directions: Determine the moisture content of various feeds provided by the instructor. Use your hand to squeeze a ball of the feed to determine an estimate of moisture content. Use the *fist test* chart as a reference to determine moisture percent.

• Stream of water	< 25% Moisture
• 6 – 10 drops	25 – 28% Moisture
• 1–5 drops	29 – 32% Moisture
• Ball doesn't open	33 – 35% Moisture
• Ball opens very slowly	36 – 40% Moisture
• Ball opens slowly	41 – 45% Moisture
• Ball opens rapidly	> 45 % Moisture

Name_____ Date_____

1. Feed description: _____
   Estimated moisture content: _____
   _____

2. Feed description: _____
   Estimated moisture content: _____
   _____

3. Feed description: _____
   Estimated moisture content: _____
   _____

4. Feed description: _____
   Estimated moisture content: _____
   _____

5. Feed description: _____
   Estimated moisture content: _____
   _____

6. Feed description: _____
   Estimated moisture content: _____
   _____

7. Feed description: _____
   Estimated moisture content: _____
   _____

Name_____ Date_____

CHAPTER 41 MATCHING ACTIVITY

Term

____ 1. challenge feeding
____ 2. transponders
____ 3. double cropping
____ 4. body condition score

Definition

a. a magnetic or electronic device attached to each cow allowing access to feed

b. refers to the amount of fat the animal is carrying

c. the practice of feeding higher levels of concentrate to challenge the cow to reach her maximum potential milk production

d. growing two crops on the same ground in the same year

Name_____ Date_____

CHAPTER 41 LAB QUESTIONS

1. What are the main goals of a dairy cow feeding program?
   _____
   _____
   _____
   _____

2. What are the common ingredients used in dairy cow feeding rations?
   _____
   _____
   _____
   _____

3. What is a TMR?
   _____
   _____
   _____
   _____

4. What procedures are essential to keep dairy calves alive after they are born?
   _____
   _____
   _____
   _____

Chapter 42
Management of the Dairy Herd

INTRODUCTION

Dairy herd and production records are the key to successful management and profitable production of milk. Dairy herd improvement associations (DHIAs) provide many records that help the dairy farmer increase net profits. DHIA records may add value to cattle and give producers a reference for improving or maintaining herd production levels. Dairy production records are standardized to a 305-day, twice daily, mature equivalent basis for comparison purposes. Several computer programs exist to help a dairy farmer properly manage the herd according to performance records.

Many modern dairy systems use microchip tags on the cattle that are read by a computer sensor. When the chip is scanned by the computer, information from each day is loaded onto the microchip, which stores all of the information and can be used to track the performance of dairy cattle more efficiently. As a dairy cow ages and production begins to slow down, herd records can track the decrease in production and allow a herd manager to make a decision of when to cull the animal from the herd. A regular culling program should be followed. Culling dairy cows is an easier decision when the price of beef is high because the dairy farmer captures more of the value of the cull cow.

It is most desirable for a dairy farm to raise its own replacement heifers. Not buying replacements can reduce the risk of introducing a new sickness to the herd. Herds that grow based on their own herd reproduction are called closed herds. Select for economically important traits when breeding cows for replacements. A proper breeding program should be established using a sire and cow that have a high genetic value and transmitting ability. Artificial insemination is commonly used on dairy farms because of the control over breeding schedule, the conception rates, and the ability to choose high-quality sires at ease, when using a semen straw. Catching the standing heat period in cows is essential for the artificial breeding program. Heifers being bred for the first time should be scheduled to calve at 22 to 24 months of age. A cow should be re-bred 50 to 60 days after giving birth.

Complete the record keeping activity and the cull cow selection lab to further your knowledge in dairy cattle management.

Farmer evaluating herd data.

Name_____ Date_____

EXERCISE 42-1 DAIRY HERD RECORD KEEPING ACTIVITY

The instructor will provide a sample farm to visit for a complete farm business analysis. The class will split into groups, and each group will be assigned a type of farm record to analyze.

Directions: Study the sample farm records for the area of data assigned. (1) Take notes on the data, (2) prepare a written analysis of the current status of the farm in this area, and (3) prepare written suggestions for improvement based on the analysis of records.

- Production records on individual cows and on the herd
- Feed use records
- Breeding and calving records
- Health records
- Cow identification records
- Financial records (the entire farm and enterprise)
- Inventory

1. **Notes**

_____
_____
_____
_____

2. **Analysis of records**

_____
_____
_____
_____

3. **Suggestions**

_____
_____
_____
_____

Name_____ Date_____

EXERCISE 42-2 CULL COW SELECTION LAB

A dairy farmer is milking 3500 head of cattle. With such a large number of cattle, the management of adding replacement heifers and removing cull cows for the herd is a full-time chore. A new crop of 500 replacements heifers are ready to be brought into the herd, and the herd manager has identified 480 cows to be culled for various reasons.

Directions: Based on the average percentage of culling due to various reasons (Table 42-1 from the textbook), determine how many cows were culled due to each reason.

Reason	Number of cows to be culled
Low production	_____
Reproduction problems	_____
Mastitis	_____
Disease	_____
Teat or udder injury	_____
Poor udder conformation	_____
Accidents and injury	_____
Poor feet and legs	_____
Other poor conformation	_____

Name_____ Date_____

CHAPTER 42　MATCHING ACTIVITY

Term

____ 1. estimated relative producing ability (ERPA)

____ 2. commercial heat detector

____ 3. net income

____ 4. gross income

____ 5. dairy herd improvement association (DHIA)

Definition

a. the total of all income received

b. income (profit) left after expenses have been deducted

c. a special halter with a cone-shaped unit containing a stainless steel ball bearing attached to the underside of the halter

d. group of dairy farmers in the local area that provides production testing services

e. a prediction of 305-2 X ME production compared to other cows in the herd

Name_____ Date_____

CHAPTER 42 LAB QUESTIONS

1. What are the main goals of a dairy producer?
   _____
   _____
   _____
   _____

2. What are the major reasons to keep accurate dairy production records?
   _____
   _____
   _____
   _____

3. Describe the practice of dehorning. Is it ethical? Why or why not?
   _____
   _____
   _____
   _____

4. Describe the tools needed for hoof trimming.
   _____
   _____
   _____
   _____

5. Describe the process of caring for a newborn dairy calf.
   _____
   _____
   _____
   _____

Chapter 43
Milking Management

INTRODUCTION

One of the most important parts of the cow is the udder. It is responsible for milk production and profitability; therefore, understanding the functions of the udder is important for dairy farmers. The udder is divided into four quarters. Each quarter has a teat that provides an outlet for the milk to be released from the udder. The alveoli in the udder wall are the component that manufactures milk. Milk letdown is the reaction of a cow to release stored milk. Gentle washing of the udder stimulates milk letdown. A hormone called oxytocin is also used to induce milk letdown.

Cows are creatures of habit, so developing a milking routine that all the milking employees follow is important for maintaining a high level of milk production per cow. Cows should be comfortable at all times. Stress on dairy cattle will cause them to let down less milk affecting the profitability of the farm. A cow should not be under- or overmilked; either way can cause discomfort and susceptibility to infection.

In order to produce high-quality, top-grade milk, several factors must be controlled. Milk quality can be maintained by keeping cows and facilities clean, rapidly cooling the milk, and properly cleaning the milking equipment. If these practices are not completed up to standards, bacteria can contaminate the milk and lower the quality of the milk. Good management practices help reduce bacteria in the milk. The level of bacteria can be measured by the presence of somatic cells. The somatic cell count of a dairy herd is a good indication of milk quality, herd health, and sanitary practices.

Milking equipment can become lined with mineral and milk stone deposits. Several disinfectants are used to remove the deposits such as alkaline, chlorinated alkaline, and acid cleaners. There are several sanitation factors that affect the quality of milk. If the milk has any off flavors or odors, the milk quality is reduced, as well as the profitability for the farmer. Recognizing off flavors and identifying the cause is an important skill to reduce quality loss.

Complete the following milk quality investigation to become familiar with the FFA Milk Quality Career Development Event.

Name_____ Date_____

EXERCISE 43-1 MILK QUALITY INVESTIGATION

In this exercise, the class should be split into teams to practice the Dairy Foods Team Event of the FFA Career Development Event (CDE). The Guide to Problem Solution is a helpful fact sheet. It is available, along with other information, on the FFA website, www.ffa.org. Look for the CDE description of milk quality and products. The guidelines are located through the link to *team activity guide*.

Directions: Use this guideline when solving the following example. Fill in the blanks in the accompanying CDE form, and then prepare a report with reasons for the evaluation that will be presented to a group of peer evaluators.

SAMPLE PROBLEM DAIRY FOODS CDE TEAM EVENT

Teams will be given a table showing results of monthly tests performed on a single dairy farm and will perform four tests to complete the table. They will use the completed table to decide whether actions should be taken for violations of any or all of the regulations or industry standards.

Below is a sample table. Teams need to fill in the blanks then prepare their report for presentation to a team of evaluators.

Test results for dairy farm #335522

	Month				
Test	1	2	3	4	5
Bacteria Count $\times\ 10^3$	20	120	45	350	____
Somatic Cell Count $\times\ 10^3$	100	500	650	750	____
Temperature (°C)	34	36	35	50	37
Antibiotic test (+/−)	+	−	−	−	____
Freezing Point (°C)	−0.522	−0.500	−0.523	−0.493	−0.529
Titratable Acidity (%)	0.18	0.16	−0.21	−0.55	____

Name_____ Date_____

Record your teams' report here:

Name_____ Date_____

CHAPTER 43 MATCHING ACTIVITY

Term

____ 1. alkaline cleaners
____ 2. acid cleaners
____ 3. off flavors

Definition

a. includes descriptions such as rancid, salty, flat, unclean, barny, and oxidized
b. may be used to remove deposits from milking equipment
c. remove inorganic deposits

Name_____ Date_____

CHAPTER 43 LAB QUESTIONS

1. Explain how a cow's udder produces milk.
   _____
   _____
   _____
   _____

2. What are some key factors to reduce stress for cows during milking?
   _____
   _____
   _____
   _____

3. How is the quality of milk assessed?
   _____
   _____
   _____
   _____

4. What factors can reduce the quality of milk?
   _____
   _____
   _____
   _____

Chapter 44

Dairy Herd Health

INTRODUCTION

The health of the cattle on a dairy farm is the most important concern for a farmer. If a cattle are affected by a sickness, the profitability of the farm decreases. If a sickness goes untreated, death of the animal is often the result. Carefully monitoring the health of a dairy herd and properly diagnosing and treating sickness must be a priority on a dairy farm. A veterinarian should be involved in all the areas of herd health when necessary. Many commercial dairy farms hire a herd health manager who focuses completely on monitoring and treating herd health issues.

A herd health plan should be developed and implemented in order to maintain a healthy, profitable dairy herd. An effective herd health plan emphasizes prevention of problems through vaccination and good management practices. Regular vaccinations and testing for common infections and diseases should be a scheduled part of a herd health program. A major piece of the herd health plan is keeping accurate records on each individual animal in the herd. These records should include health history, record of treatments, milk production records, and reproduction records. The individual animal records are very helpful in accurately diagnosing and treating health problems.

Dairy cattle are at risk of infection from several diseases, bacteria, and parasites. A herd health manager must understand the symptoms and treatments necessary to control the impacts of the infection and reduce the amount of profit loss. Some common health complications include displaced abomasum, retained placenta, ketosis, metritis, and milk fever, as well as internal and external parasites. Mastitis is the most serious disease that affects dairy cattle because it directly affects the milk quality. Mastitis can be tested, and a measure of somatic cells present will indicate the level of mastitis infection in a herd of milking dairy cows. The average somatic cell count in a healthy dairy herd should be below 200,000.

Careful management and proper treatment of all heard health problems is needed to keep losses low. Using medication properly and allowing for withdrawal times is an important aspect of herd health. In this chapter, a herd health plan will be developed for a sample dairy herd.

Name_____ Date_____

EXERCISE 44-1 DEVELOP A HERD HEALTH PLAN LAB

Directions: Various herd health management areas are listed in this exercise. For each area, describe the practices that should be followed to diagnose, treat, and reduce the spread of disease and parasites.

Herd Health Plan

Farm name: _____

Herd health manager name: _____

1. General list of herd health practices that help reduce sickness. All employees should follow these practices when handling the health of dairy cattle. (List 8):

   _____
   _____
   _____

2. Biosecurity measures on the farm

Measures Taken with Incoming Animals	Practices Followed on Farm
• Sourcing policy	
• Ascertaining disease status	
• Pre-movement/purchase information from seller	
• Avoid/minimise mixing of stock	
Nominated Isolation Facility/Area	
• Location of nominated Isolation facility or area	
• Means of cleaning and disinfecting after use	
Other Measures	
• People	
• Buildings	
• Equipment	
• Vehicles	
• Fencing	

Name_____ Date_____

3. Infectious disease control (information found in Chapter 17 of the textbook)

Disease	Prevention/Treatment (vaccines)/Control Measures/Timing
Bruccelosis	
IBR	
BVD	
Parainfluenza-3	

4. Parasite control

Parasite	Prevention/Treatment(vaccines)/Control Measures/Timing
Fluke	
Lungworm	
Intestinal worm	
Lice	
Fleas	
Ticks	

5. Mastitis

Condition	Measures Taken	Product Used	Withdrawal Period
Environmental mastitis			
Contagious mastitis			

Name_____ Date_____

6. Other disease and health management problems

Condition	Prevention/Treatment (vaccines)/Control Measures/Timing
Milk fever	
Ketosis	
Displaced abomasum	
Retained placenta	

CHAPTER 44 MATCHING ACTIVITY

Term

____ 1. leukocytes
____ 2. somatic cells
____ 3. placenta
____ 4. mastitis

Definition

a. involuntary muscle contractions
b. the most serious disease that affects dairy cattle
c. white blood cells that fight infection
d. the afterbirth of a cow

Name_____ Date_____

CHAPTER 44 LAB QUESTIONS

1. Why is herd health one of the most important areas of farm management?
   _____
   _____
   _____
   _____

2. What are some important steps to controlling infections of diseases and parasites?
   _____
   _____
   _____
   _____

3. List the most common diseases of dairy cattle.
   _____
   _____
   _____
   _____

4. Describe the common test used to detect mastitis. What is the proper treatment if a cow is infected?
   _____
   _____
   _____
   _____

Chapter 45

Dairy Housing and Equipment

INTRODUCTION

Dairy farms are very high in capital expenses. Some of the areas of the business that require investment are animal housing, milking facilities, manure handling facilities, labor expenses, and feed production and handling. Well-planned facilities save labor and make the operation more efficient in many ways. When designing a facility, the producer should address issues such as the location of the facility, the size of the planned herd, the laws and regulations that apply to dairy farms, the source and amount of money available, the type of milk market available, the amount of labor available, the kind of housing system to be used, the type of milking system to be used, the feed handling system to be used, the manure handling system to be used, and any future expansions that might be desired. Each of these areas must be thought of during the planning process; the goal should be efficiency, economy, and convenience.

The two most common types of housing used for the milking herds in the United States are stall (stanchion) barns and free stall barns. Most dairy farms that are milking more than 80 cows use free stall barns due to their efficiency. Many small dairy farms use tie-stall or stanchion barns. Many times, these buildings already existed; it is unusual for a modern dairy farm to build a new stanchion barn. Farmers that use a stanchion barn have a more labor-intensive milking system. Pail milkers, suspension milkers, and pipeline milkers are used in stall barns. Milking parlors with pipeline milkers may also be used in stanchion barns.

Fee stall dairy barns are usually paired with a milking parlor. The most common type of milking parlor is the herringbone. Other types are side-opening, rotary or carousel, and polygon. Many free stalls are equipped with automated equipment. This mechanization and automation requires a large capital investment, but the reduction in labor and increase in efficiency is proven to pay off when managed correctly. The use of crowding gates, power gates, feed gates, stimulating wash sprays, and automatic detaching units are just some examples of the automation available for modern milking parlors.

Between the free stall barn and the milking parlor, most designs include a concrete holding area to funnel cows into the milking parlor and contain them while they wait to be milked. The crowd gate is usually located in the holding area and helps push the herd through the parlor. Attached to the parlor is where a milk house is often located. A milk house contains the equipment for filtering, cooling, and storing the milk. A milk house in inspected by the milk inspector and must pass sanitary regulations. Large bulk tanks are located inside or just outside of the milk house. The bulk tanks are refrigerated and stored milk at the proper temperature. With all of the mechanization of modern dairy farming systems, the energy costs can be very high. Good management practices reduce the energy bill by eliminating waste and exploring alternative energy sources.

The milking system uses a good portion of the total energy requirements of a dairy farm. It has several systems that must work properly to retrieve milk, cool it, and send it to the bulk tank for storage. A milking system is made up of the milking unit, the pulsation system, the vacuum supply system, and the milk flow system.

The calves and the older heifers on a dairy farm may be housed in buildings that require less capital investment. Either system works well with good management. Housing must be clean, dry, well ventilated, and free from drafts. Older heifers may be grouped in pens or kept in free stall barns.

Manure handling on a modern dairy farm is highly mechanized. Most manure is in the liquid form. It may be stored in a lagoon until being spread as fertilizer on crop fields. Some farms have begun using methane digesters as a way to utilize the energy from the waste products.

Modern carousel milking parlor.

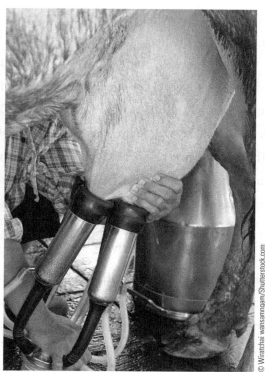

Stanchion barn milking system.

Name_____ Date_____

EXERCISE 45-1   DESIGN A MODERN CATTLE HANDLING FACILITY LAB

Two examples of dairy farms are provided here. Each description gives information about the size of the herd and systems that are used.

Directions: Read each dairy farm description. For each farm, draw an aerial view map of the facilities. Include each component of the farm in the drawing (that is, type of barn, milk house, parlor, manure handling, etc.). Be sure to label each building or structure. Use straight lines.

Example Dairy Farm 1

This farm has 1100 head of cattle—600 are milking cows, 200 are calves, 200 are heifers, and 75 are dry cows, and there are 25 bulls on the farm that will be marketed for genetics. There are two barns used for the milking cows. Each barn has 275 stalls; there are fans located through out the barn, misters to keep cows cool in the summer, large waterers throughout several areas of the barn, and an automatice scraping blade that slowly pushes manure into a gutter the end of the barn. The manure is then pumped from the gutter to a lagoon, where it is stored until there is time to spread the manure on the fields.

When it is milking time, the cows are moved from the free stall to the milking parlor and contained in a holding area while they are waiting to be milked. The milking parlor is a double 20 parallel parlor with automatic takeoff milking units. The milk is pumped to three separate 3000-gal milk tanks, where it is stored until the milk truck comes. The tanks are located just outside the milk house. The milk house contains a steam cleaner and sinks to keep equipment sanitary. The farm office is also located in the milk house. The calves are raised in calf hutches until they are 3 months old. From that time until they are lactating, they are housed in smaller confinement barns with a bedded pack. Each barn on the farm has a separate pen for sick animals that need treatment.

Example Dairy Farm 2

This is a small dairy farm with 110 head of cattle—75 are milking cows, and the other 35 animals are calves, heifers, and dry cows. A stanchion barn is used for the milk cows. The milking operator brings a milking unit to each cow and then attaches the unit to the vacuum and milk lines that run the length of the barn. There are two rows of cows facing away from each other with feed mangers in front of the cows and a manure gutter located behind them. Inside the manure gutter is a large chain run by a motor that moves the manure out of the barn and into a manure spreader that must be emptied onto the fields each day. The milk house is attached to the barn. It contains a sink to clean equipment and a 1500-gal milk tank in the center of the milk house.

Calves are located in a small building attached to the back side of the stanchion barn. It has small box stalls for young calves and larger pens with a bedded pack for older heifers. There is also a box stall for sick cows. The farm office is located in the farmer's house.

Name_____ Date_____

Locate both design drawings in the space given below (include all needed facilities):

Name_____ Date_____

CHAPTER 45 LAB QUESTIONS

1. What the major facility systems are needed on a dairy farm?
   _____
   _____
   _____
   _____

2. What type of dairy operation is more efficient? Which is more economical?
   _____
   _____
   _____
   _____

3. List the various types of milking systems and parlors.
   _____
   _____
   _____
   _____

4. Describe the types of manure handling systems.
   _____
   _____
   _____
   _____

5. What type of housing is most commonly used for raising calves?
   _____
   _____
   _____
   _____

Chapter 46
Marketing Milk

INTRODUCTION

The dairy market is constantly fluctuating throughout the year. Long-term trends have made it possible to predict when the price paid for milk will rise and drop. The long-term data show a seasonal trend in the dairy market. The dairy production levels compliment the levels of milk price; the price of milk is usually higher when there is less production. The highest production occurs during the spring, and the lowest prices are paid in the 6 months following the spring.

Milk used for fluid product, cheese, butter, and ice cream make up about 96 percent of the total milk processed every year in the United States. The consumption rate of dairy products has shown a decreasing trend since the 1980s with the exception of cheese. Dairy advertising and education have been highly funded by promotional programs with the purpose of reversing this trend. If there is more milk on the market than demand, there are several protection programs in place to take product off the market in order to increase price paid. An example of this strategy was the dairy herd buyout of the 1980s. This program took action to save a struggling dairy industry by paying farmers to sell their entire herd for slaughtering. This took many cows out of production and lowered the supply of milk significantly. No actions this serious have been necessary since that time.

About 86 percent of the milk produced in the United States is marketed through cooperatives. Cooperatives provide many services to milk producers as well as to the entire market. Most of the cooperatives provide trucking of the milk from the farm to the processing plant. The cost of transportation has risen greatly over recent years due to rising fuel and labor costs. A fee for trucking is usually taken out of the farmer's milk check automatically.

Milk is divided into classes based on use for pricing purposes. Federal milk marketing orders are established in many areas to set prices for Grade A milk. The actual price paid is often above the minimum set in the federal order market.

Name_____ Date_____

EXERCISE 46-1 DAIRY ADVERTISING ACTIVITY

The dairy industry is known for its advertising campaigns. Dairy cooperatives and milk promotion agencies direct a lot of funding to produce effective advertising campaigns. The famous "Got Milk" slogan is known by most Americans. A simple advertisement poster with a picture of a celebrity drinking a glass of milk and wearing a milk mustache has been very effective. In this activity, the class will be split into advertising teams. Each team will need to complete Parts 1 and 2. Prior to completing Parts 1 and 2, review several TV advertisements for milk to become familiar with common dairy commercials.

Part 1

Directions: Create an advertising poster that can be placed in grocery stores, in school cafeterias, on billboards, in magazines, and with other types of advertising media. The poster should have some type of slogan to identify the objective of the advertisement and visuals that immediately make consumers think about milk products. Hand the poster into the instructor for grading; then find a place in the school where the poster could be displayed. Posters should be completed on separate paper material and handed into instructor.

Name_____ Date_____

Part 2

Directions: The instructor has provided several examples of dairy products. Create a TV advertisement lasting approximately 30 seconds to promote the product that the group chose. Record the advertisement to be shown to the class. Use the following form to prepare and plan for the advertisement.

Type of product	
Name of product	
Slogan	
Music	
Type of advertisement (funny, romantic, etc.)	
Script: conversation/voiceover	

Name_____ Date_____

EXERCISE 46-2 MILK GRADING LAB

Background: Milk is classified into different types based upon the amount of fat. Skim milk contains 0 g of fat per 250-ml serving; 2% milk contains 2 g of fat per 250-ml serving, whole milk contains 8 g of fat per 250-ml serving; half and half cream contains 8 g of fat in a 2-Tbsp serving. Homogenization of milk breaks up the fat into very small fat globules and spreads them throughout the milk. The fat globules (0.1 to 15 μm in diameter) are basically suspended uniformly in the milk. Due to the fat content of milk, when it is mixed with a small amount of dish soap and food coloring, the milk begins to move and flow. The movement can be observed to determine which class of milk the sample belongs in.

Directions: Select four milk products from the following list. Create a hypothesis, follow the lab procedures, and record results in the data table; then conclude the lab based on the results.

Materials

Petri dishes	Liquid dish detergent	Toothpicks
Watch glasses	Assorted food colorings	Skim (fat-free) milk
1% milk	2% milk	Whole milk
Half and half	Buttermilk	Goat's milk
Powdered milk		

Hypothesis

Rank the milks for what you believe will be the least to most movement after adding the food coloring. Provide an explanation (rationale) for your predictions.

Procedures

1. Place the four Petri dishes on the lab bench. Use a small piece of paper to label each dish.
2. Choose four of the eight milk choices provided for the lab.
3. Fill each Petri dish half full with the appropriate milk.
4. Equally space four drops of food coloring in each dish. Record your observations of the activity (movement) of the food coloring in the milk.
5. Dip a toothpick into the liquid dishwashing detergent. Then, touch and hold the toothpick into the middle of each dish.
6. Try again with more detergent touching the milk in different areas. Record your observations of the activity (milk/food coloring movement) when soap is added.
7. Pour the used milk down the drain with lots of water. Clean each Petri dish with lots of soap and water; then dry with paper towels. Return all the materials and clean the lab area.

Name_____ Date_____

Data Table

Type of Milk	Observations	Rank (least to most movement)

Conclusion

Write at least one paragraph describing the results of the lab and explaining the data. This should compare the hypothesis to what actually was observed during the experiment.

_____
_____
_____
_____
_____
_____
_____

Name_____ Date_____

CHAPTER 46 LAB QUESTIONS

1. Describe the yearly trends for dairy production and pricing.

   _____
   _____
   _____
   _____
   _____

2. What has been the general trend for dairy consumption since the 1980s?

   _____
   _____
   _____
   _____
   _____

3. Describe some dairy promotion programs and how they promote increased consumption.

   _____
   _____
   _____
   _____
   _____

4. What is the difference between Grade A and Grade B milk?

   _____
   _____
   _____
   _____
   _____

SECTION 10

Alternative Animals

Chapter 47	Rabbits	340
Chapter 48	Bison, Ratites, Llamas, Alpacas, and Elk	348

Chapter 47

Rabbits

INTRODUCTION

The rabbit is sometimes overlooked as a production animal. However, rabbits are raised through production agriculture for purposes such a meat production, pets/companionship, and scientific research. Modern breeds of rabbits have been developed and modified from European wild rabbits. The major use of domestic rabbits is for meat production. Rabbits produce a white meat that is high in protein and low in calories, fat, and cholesterol. The most common meat breed is the New Zealand white breed.

A feeding program is just as important in rabbit production as it is with any other type of livestock. Rabbits have a simple stomach, so plant products make up the majority of their diet. Most commercial rabbit producers as well as pet owners feed a completed pellet. There is not usually a need for additives or supplements in the pelleted ration. Just like other livestock, rabbit require fresh, clean water. Rabbits that are growing or used for reproduction have a higher energy requirement, so they are fed a completed pellet and have access to legume hay. Rabbits that are on a maintenance diet only require hay in their diet.

One of the advantages of the rabbit production industry is the short gestation periods of rabbits. The gestation period for rabbits is 30 to 32 days. Does that are in good physical condition can be rebred 6 weeks after kindling. On an accelerated breeding program, a doe can produce up to five litters per year.

Rabbits are difficult to treat for disease; usually when symptoms are observed, treatment cannot save the animal. Good sanitation in the rabbitry is the key to disease control. Keep a clean environment, and reduce the amount of visitors to prevent disease transmission. Keep dogs, cats, birds, rodents, and insects out the rabbitry because they sometimes carry diseases or parasites that may be transmitted to the rabbits.

The exercises in this chapter reinforce an understanding of the rabbit breeds and identification of rabbit health problems.

Production rabbits.

342 SECTION 10 Alternative Animals

Name _____ Date _____

EXERCISE 47-1 BREED SELECTION ACTIVITY

There are 47 breeds of rabbits according to the American Rabbit Breeders Association (ARBA) Standard of Perfection. Each breed has specific characteristics such as size, color patterns, physical attributes, and purposes.

Directions: For each of the following breeds of rabbit, research and record the breed characteristics and functions. In the square provided, paste a picture of the breed. These pictures may be cut out of magazines or found on the Internet and printed from a computer.

1. Checkered Giant
   a. Description _____
   _____
   b. Function _____
   _____
   c. Picture _____
   _____

2. Dutch
   a. Description _____
   _____
   b. Function _____
   _____
   c. Picture _____
   _____

© 2016 Cengage Learning®. May not be scanned, copied or duplicated, or posted to a publicly accessible website, in whole or in part.

Name_____ Date_____

3. New Zealand
   a. Description _____
   _____
   b. Function _____
   _____
   c. Picture _____
   _____

4. California
   a. Description _____
   _____
   b. Function _____
   _____
   c. Picture _____
   _____

5. English Spot
   a. Description _____
   _____
   b. Function _____
   _____
   c. Picture _____
   _____

344 SECTION 10 Alternative Animals

Name_____ Date_____

EXERCISE 47-2 RABBIT HEALTH PROBLEM LAB

There are many health problems that can affect rabbits. If quick diagnosis and treatment does not happen, death of the animal is often the result. Part 1 will practice matching symptoms with the health problems; in Part 2, students will perform physical examinations of rabbits looking for signs of health problems.

Part 1

Directions: Several examples of symptoms due to various rabbit illnesses are listed here. For each description of symptoms, (a) identify the health problem, (b) what caused the problem, and (c) the possible treatments.

1. Diarrhea containing blood. Slow to no weight gain or possible loss of weight. May be potbellied after recovery.

    a. _____
    b. _____
    c. _____

2. Loss of appetite; loss of weight. Diarrhea may be exhibited. Fur is rough. Pneumonia may develop.

    a. _____
    b. _____
    c. _____

3. Breasts become swollen and feverish; turn black and purple. Abscesses may form. Animal has fever.

    a. _____
    b. _____
    c. _____

4. Hair loss in circular patches; scaly skin with red raised crust. Matted fur. May occur on any part of the body.

    a. _____
    b. _____
    c. _____

5. Infected, inflamed, bruised, or abscessed bare areas on hind legs. Animal shifts weight to front legs. May affect front legs in severe cases. Secondary infections may occur.

    a. _____
    b. _____
    c. _____

Name_____ Date_____

6. External genitals and anus become inflamed. Crusts may form; bleeding may occur. Pus may develop in severe cases.

   a. _____
   b. _____
   c. _____

7. Dry, scaly, irritated skin, scratching, loss of fur on head, ears, and neck.

   a. _____
   b. _____
   c. _____

Part 2

Directions: The instructor will provide several sample rabbits. Give each rabbit a physical and visual examination. Palpate the body and pay attention to areas commonly affected by known health problems. If any health problems are detected, attempt to diagnose the problem and then report the problem to the instructor.

Name_____ Date_____

CHAPTER 47 MATCHING ACTIVITY

Term

____ 1. herbivorous
____ 2. coprophagy
____ 3. kindling
____ 4. pseudopregnancy
____ 5. palpate
____ 6. hutch

Definition

a. refers to moving the hand gently back and forth, exerting a slight pressure
b. false pregnancy
c. refers to the ingestion of fecal matter
d. animal shelter/housing
e. giving birth
f. their diet comes mainly from plant sources

Name_____ Date_____

CHAPTER 47 LAB QUESTIONS

1. Where did today's modern rabbit breeds descend from?

   _____
   _____
   _____
   _____

2. What breed is the most popular for meat production?

   _____
   _____
   _____
   _____

3. What type of feed makes up a rabbit's diet?

   _____
   _____
   _____
   _____

4. How many gestations can a rabbit have every year?

   _____
   _____
   _____
   _____

Chapter 48

Bison, Ratites, Llamas, Alpacas, and Elk

INTRODUCTION

Bison are an unforgettable part of U.S. history. The American bison once numbered in the millions on the North American continent and were an important resource for the Plains Indians. The settlers of the United States hunted the bison with no regulation or control, and the population was nearly brought to extinction. Wild and domestic populations of bison are on the rise. A growing number of people are raising bison for sale as breeding stock and for meat. Bison production takes a great deal of knowledge and care as well as safe animal handling practices. Bison are a very large, strong animal that is less tame than domestic breeds of cattle, so extra precaution must be taken when working with the animals. Bison need much less investment in housing and equipment than most domestic cattle, and they can grow on a similar feed ration as most domestic cattle.

Ratites are flightless birds and include ostrich, emu, rhea, cassowary, and kiwi. The ratite industry in the United States has been growing, and the most common ratites for production are the ostrich and emu. These two species are the largest birds still in existence on earth, and they are fairly efficient converters of feed. The ostrich hen will usually produce 40 to 50 eggs per year, and the emu hen will produce about 30 eggs each year. Ratites are fed rations similar to poultry, but not enough research has been done to fully understand the nutrient requirements of ratites. One reason that more livestock producers do not raise ratites is that some of them can be very dangerous to handle and are aggressive toward humans. Without much research into ratite management, the best control of disease is prevention. There is not much medicine on the market specifically for ratites, and the population of veterinarians who will handle ratite cases is small.

Llamas and alpacas have a rich history in the Andes Mountains of South America, dating back 4000 or 5000 years ago. Llamas are used as pack animals as well as for meat, hides, and wool. Being smaller than llamas, alpacas are not usually used as a pack animal but are raised for their wool, which has softer fibers than llama or sheep wool. Llamas and alpacas are modified ruminants, having three stomachs. The feeding program is based on the use of roughages, but the use of commercially prepared pelleted feeds is common. Llamas and alpacas are similar to horses in the fact that they are very trainable and usually respond well to humans. Neither animal has an estrus period; rather they can be induced into ovulation and bred during any time of the year. Herd health concerns are minimal for llamas and alpacas, but, as with any livestock animal, good sanitation and feeding programs will reduce health problems.

There are a growing number of American Elk farms across the country. Elk take very specialized handling practices and equipment. A producer must have experience and be very knowledgeable about the behaviors of elk. They are primarily raised for their velvet, which may produce 40 to 50 lb per year. Many Asian countries use the velvet for medicinal remedies. Elk are also bred for meat, by-products, and breeding stock.

Bison on fenced in pasture.

Name_____ Date_____

EXERCISE 48-1 BISON MANAGEMENT LAB

Bison are very similar to cattle in their feed requirements and health concerns. Despite the similarities, bison are a less domesticated animal; they are larger, stronger, and faster than cattle. These traits make herd management more difficult—and potentially dangerous. Specific handling and management tasks should be planed according to the characteristics of bison.

Part 1

Directions: For each of the following areas of livestock management, identify three differences in management practices compared with other types of livestock.

Animal handling

1. _____
2. _____
3. _____

Breeding

1. _____
2. _____
3. _____

Weaning

1. _____
2. _____
3. _____

Animal identification

1. _____
2. _____
3. _____

Dehorning

1. _____
2. _____
3. _____

Name_____ Date_____

Herd health

1. _____
2. _____
3. _____

Feeding practices

1. _____
2. _____
3. _____

Facilities and equipment

1. _____
2. _____
3. _____

Part 2

Directions: Read the following example of a feeding and water program for a small herd of bison. Answer the calculation questions using the information provided in the example.

Example: A bison will usually drink 5 gal of water each day. In this particular herd, each bison eats 3 lb of feed cubes, three times a week. This bison farmer feeds two 800-lb round bales to the herd each week. There are 68 head of bison in this herd.

1. How many round bales are needed for the entire year? _____
2. How many pounds of cubes does one bison eat in 2 weeks? _____
3. How much water does the entire herd drink in 1 day? _____
4. How many gallons of water does one bison need for a week? _____
5. How many pounds of cubes will be needed to feed the herd for 1 week? _____
6. One water trough holds about 80 gal of water. If it fills up only once, how many water troughs will be needed to water the whole herd each day? _____

Name _____ Date _____

EXERCISE 48-2 HISTORY OF LLAMAS AND ALPACAS ACTIVITY

Directions: List at least three major historical milestones for each period during the history of llamas and alpacas.

1. **Evolutionary history**

2. **South Amercian history**

3. **History in the United States**

4. **Description of the modern llama and alpaca industry**

Name_____ Date_____

CHAPTER 48 MATCHING ACTIVITY

Term **Definition**

____ 1. bloodtyping **a.** the name for a baby llama, alpaca, or other camelidae

____ 2. ratite **b.** the male parent of an animal

____ 3. cria **c.** the female parent of an animal

____ 4. sire **d.** is used to identify individual animals and to determine parentage

____ 5. dam **e.** elk produce 30–40 lb of this each year

____ 6. velvet **f.** a group of flightless birds that include ostrich, emu, rhea, cassowary, and kiwi

Name_____ Date_____

CHAPTER 48 LAB QUESTIONS

1. What changes have occurred in the bison population throughout American history?
_____
_____
_____
_____

2. List four ratites.
_____
_____
_____
_____

3. How many eggs does an emu produce every year?
_____
_____
_____
_____

4. What was the original purpose of llamas? Where do they originate from?
_____
_____
_____
_____

References

Anderson Bruce, Terry Mader, and Paul Kononoff. *Sampling Feeds for Analyses*. University of Nebraska—Lincoln Extension, Institute for Agriculture and Natural Resources, 2007.

Barrick, Kirby R., and Harmon, Hobart L. *Animal Production and Management*. New York: McGraw-Hill, 1988.

Bundy, Clarence E., Diggins, Ronald V., and Christensen, Virgil W. *Livestock and Poultry Production*, 5th ed. Englewood Cliffs, NJ: Prentice Hall, 1982.

"Camelid Facts." *Bar Q Diamond Ranch*. 2 January 2014. B-Q-Diamond Ranch Inc. www.bar-q-diamond.com/information/llamafacts.htm.

"Cattle Handling and Working Facilities." *Beef Quality Assurance*. Ohio State University Extension, 2002.

"Chicken Wing Anatomy Laboratory." *Life Science*. K12 Inc., 2004.

Duberstein, Ph.D., Kylee J., and Edward L. Johnson. "How to Feed a Horse: Understanding Basic Principles of Horse Nutrition." *UGA Cooperative Extension Bulletin 1301*. University of Georgia Cooperative Extension, 2012.

Earl Colverson, Kathleen. *Animal Science Anatomy & Physiology Lab Manual*. Clifton Park, NY. Cengage Learning, 1998.

Fairchild, Brian D., and Casey W. Ritz. "Poultry Drinking Water Primer." *UGA Cooperative Extension Bulletin 1301*. University of Georgia Cooperative Extension, 2012.

Filley, Shelby, and Amy Peters. *Goat Nutrition Feeds & Feeding*. Oregon State University Extension Service.

Gillespie, James R. *Animal Nutrition and Feeding*. Clifton Park, NY. Cengage Learning, 1987.

Gillespie, James R. *Animal Science Lab Manual*. Clifton Park, NY. Cengage Learning, 1998.

Gillespie, James R. *Modern Livestock & Poultry Production: Seventh Edition*. Clifton Park, NY. Cengage Learning, 2004.

"Horse." *Kentucky 4-H Horse Volunteer Certification Resource Manual*.

Incubation and Embryology Project. University of Illinois Extension, 1999. N. pag. www.urbanext.uiuc.edu/eggs.

Johnson, Jason, Blake Bennett, Stan Bevers, Brenda Duckworth, Wade Polk, and Bill Thompson. Department of Agricultural Economics, Texas Cooperative Extension, Texas A&M University. September 2005.

"Leader Resources Manual." *Alberta 4-H Horse Project*.

McAvoy, Jackie. "Buy Me, Buy Me!" *Integrated Skills*. Macmillan Publishers Ltd, 2008. *Teacher's Notes*. www.onestopenglish.com/.

Parker, Rick. *Equine Science: Fourth Edition*. Clifton Park, NY. Cengage Learning, 2013.

Poultry Industry Teaching Resource. Poultry CRC. www.poultryhub.org.

Rasby, Rick, and Jeremy Martin. *Understanding Feed Analysis*.

Wagner, J., and T. L. Stanton. "Formulating Rations with the Pearson Square." *Livestock Series|Management*. Colorado State University Extension, 2012.

Ziehl, Amanda, Megan L. Bruch, Aaron C. Robinson, and Rob W. Holland Jr. *Goat Producers*.